THE THEORY OF INDISTINGUISHABLES

SYNTHESE LIBRARY

STUDIES IN EPISTEMOLOGY,

LOGIC, METHODOLOGY, AND PHILOSOPHY OF SCIENCE

VOLUME 150

A. F. PARKER-RHODES

THE THEORY OF INDISTINGUISHABLES

A Search for Explanatory Principles Below the Level of Physics

D. REIDEL PUBLISHING COMPANY

DORDRECHT : HOLLAND/BOSTON : U.S.A.

LONDON : ENGLAND

Library of Congress Cataloging in Publication Data

Parker-Rhodes, A. F. (Arthur Frederick), 1914–
The theory of indistinguishables.

(Synthese library; v. 150)
Bibliography: p.
Includes indexes.
1. Algebraic logic. 2. Combinatorial analysis.
3. Mathematical physics. I. Title.
QC20.7.A4P37 530.1'5 81–12061
ISBN-13: 978-94-009-8403-5 e-ISBN-13: 978-94-009-8401-1
DOI: 10.1007/978-94-009-8401-1

Published by D. Reidel Publishing Company,
P.O. Box 17, 3300 AA Dordrecht, Holland.

Sold and distributed in the U.S.A. and Canada
by Kluwer Boston Inc.,
190 Old Derby Street, Hingham, MA 02043, U.S.A.

In all other countries, sold and distributed
by Kluwer Academic Publishers Group,
P.O. Box 322, AH Dordrecht, Holland.

D. Reidel Publishing Company is a member of the Kluwer Group.

TABLE OF CONTENTS

PART II: APPLICATION

PREFACE AND ACKNOWLEDGEMENTS

This work has been in progress, on and off, for nearly twenty years, and during this time I have received help and encouragement from many sources. The most closely connected development to which my work is related, and to which I have made major contributions, has come to be known as the combinatorial hierarchy. One strand of this work, of which Whitehead and Eddington were precursors, was the discussion of space-time structure from an algebraic point of view, the dimensions being regarded as (we would say now) a combinatorial structure, and in fact isomorphic with the first level of the hierarchy as now known (E. W. Bastin and C. W. Kilmister, *Proc. Camb. Phil. Soc.* **50** (1954), 278). Another step was the consideration by Ted Bastin of a hierarchical switching and multiple feedback loop model in which the points in spaces were built up in a hierarchical manner with the dimension structure appearing at the simplest stage. Gordon Pask collaborated, and constructed a hardware form. In 1961 I made an algebraic formalization of the model in terms of binary variables. I invented the matrix mapping representation of the level connection and the use of matrices as the new vector operands. I discovered the breakdown of the construction at the fourth level, and associated the maximum numbers of eigenvalues with the coupling constants. This collaborative work with John Amson, Ted Bastin, and Clive Kilmister was published by Bastin (*Studia Philosophica Gandensia* **4**(1966), 77).

For some years there was disagreement on the physical interpretation of the vectors. My own thinking about the problem of interpretation centered on the question of indistinguishability, and has led to the logical structure presented in this volume. The others have followed a different approach, attempting to construct the 'particles' of physics concurrently with the 'background' that gives them meaning using only finite concatenating strings of the existence symbols '0' and '1', which both engage in elementary discriminations and construct the hierarchy sequentially. Particularly after they were joined by Pierre Noyes in the 70's, Amson, Bastin and Kilmister have placed increasing emphasis on the possible interpretation of the existence symbols as representing the presence or absence of conserved quantum numbers and on possible connections with the statistical and evolutionary paradigm of high energy particle scattering experiments. A preliminary version of their work has been published by Bastin, Noyes, Amson and Kilmister (*Int. Jour. Theor. Phys.* **18** (1979), 445); an integrated presentation of their work is under way in a volume entitled 'Combinatorial Physics' being prepared by Bastin. Since their approach is, at least superficially, quite different from my own, I am particularly grateful to them for their advice and encouragement, and the hard work they have put into the detailed criticism of my text – Noyes with regard to the physics, Bastin with regard to foundational principles, and Kilmister with regard to mathematical structure. My gratitude is deepened by the fact that their help persists in spite of their continuing disagreement with many of the ideas I present in this volume. I am also indebted to the Alternative Natural Philosophy Association founded by Noyes, Bastin and Kilmister in 1979 for help in getting my views before the public.

FOREWORD

It is widely assumed that there exist certain objects which can in no way be distinguished from each other, unless by their location in space or other reference-system. Some of these are, in a broad sense, 'empirical objects', such as electrons. Their case would seem to be similar to that of certain mathematical 'objects', such as the minimum set of manifolds defining the dimensionality of an R-space. It is therefore at first sight surprising that there exists no branch of mathematics, in which a third parity-relation, besides equality and inequality, is admitted; for this would seem to furnish an appropriate model for application to such instances as these. I hope, in this work, to show that such a mathematics in feasible, and could have useful applications if only in a limited field.

The concept of what I here call 'indistinguishability' is not unknown in logic, albeit much neglected. It is mentioned, for example, by F. P. Ramsey [1] who criticizes Whitehead and Russell [2] for defining 'identity' in such a way as to make indistinguishables identical. But, so far as I can discover, no one has made any systematic attempt to open up the territory which lies behind these ideas. What we find, on doing so, is a body of mathematics, offering only a limited prospect of practical usefulness, but which on the theoretical side presents a strong challenge to conventional ideas. It is, at the least, a fascinating curiosity; but I suspect it may go deeper than that.

I divide the work into two Parts, of which the first will be devoted to the theory, while in the second I shall explore one possible application of it. It is the second part which has, in some measure, determined the pattern of development of ideas in the first part. But it need hardly be said that the theory, as a work of mathematics, does not depend on the success of any particular application of it which may be attempted, but only on the consistency of the arguments which it contains, and the rigour with which they are presented.

However, in the matter of rigour, a general warning to the reader is in order. This is a well-defined concept, and one which calls for no modification in the context of the present work. But its practice has been, up to now, confined to systems resting on an axiomatic basis both simpler and more familiar than that which will be presented here. It is therefore a bold claim to have maintained as high a standard of rigour in this work as mathematicians customarily expect; and I am conscious that I may not always have achieved it. As the reader will soon discover, it is not easy to check the validity of particular inferences, on account of the pitfalls occasioned by the presence of three parity-relations. The matter can be shown to be in general decidable; but it would be strange indeed if my work should in the end achieve more than historical interest, as the first attempt in a new field. It is indeed to be hoped that it will be followed by exposition both more rigorous and more lucid.

PART I

THEORY

Difficulties of exposition make themselves felt from the first, and it is not hard to understand why the idea of indistinguishables has been so long neglected. Quite apart from the philosophical problems (which I shall not go into) these difficulties include not only, as is inevitable, a revised and more complex axiom-schema to replace the familiar rules about the substitutability of equals, but also a whole preliminary section of analysis which can ordinarily be passed over in silence. This concerns the semantics of mathematical notations. Every statement made thereafter needs to be scrutinized more laboriously than one ordinarily expects, if one is to assure oneself that the argument does indeed hold up. The reader who wishes to check the validity of my reasoning must thus work harder than usual, and I can only hope that the interest and novelty of the exercise may furnish an adequate reward.

CHAPTER I

INTRODUCTION

In view of the afore-mentioned difficulties, it will be helpful to begin with a preview of the argument. I start, in this chapter, by formulating a pre-mathematical characterization of the concept of 'indistinguishability'. This will, of course, have to be reformulated later in mathematically rigorous terms. But that will not become possible, till we have made a thorough semantic analysis of the kinds of statements that can be made in the presence of this unfamiliar relation. This is not a normal part of any mathematical theory, but if the criteria of argument are to be securely grasped, it must be gone into with as much care as the rest. This will form the material of Chapter II, and its end product will be an axiom-schema which will replace the conventional one concerning the conditions under which one symbol may be validly replaced by another. The change in this rule (which normally goes unstated because so much is taken for granted) to a more complex one, constitutes the main trauma of this theory.

In Chapter III I shall discuss the problem of whether theories involving indistinguishables can be correlated convincingly with empirical data, and, if so, how this should be done. This topic is introduced thus early because, if its feasible solution is not clarified, the reader may be ill-prepared to carry on to the end of this part. We shall find that in this matter, as in several other places, the logical status of the theory is similar to that of conventional mathematics, though with an extra twist not normally met with. It will appear that valid applications can be made, but subject to certain conditions which will be seen later to have curious consequences.

The next two chapters, IV & V, contain the groundwork of the theory; they have one advantage, that they are not very long. Indeed, the actual content of this theory is much poorer, as one might expect, than that of conventional mathematics. Especially the theory of mappings, developed in Chapter V, is chiefly characterized by the many areas which the third parity-relation reduces to triviality. Here then the going is hard, but the way is short; though there remain many possible lines of advance which I don't pursue, since they have no apparent relevance for the purposes of Part II. In Chapter VI I return to the conditions for empirical relevance set out in Chapter III, and ask how they are to be met; it will appear then that the classes of indistinguishables which can be rationally correlated with data are not many. Few enough, indeed, for me to attempt in Chapter VII to construct 'representations' of all of them. These representations will be put to use in Part II, to which the over-trustful may now be tempted to turn.

1.A. THE CONCEPT OF INDISTINGUISHABILITY

I start, for simplicity's sake, from current ideas about subatomic particles. Two electrons are conventionally thought of as being 'equivalent' in the mathematical sense. Two objects, whether material or mathematical, are said to be *equivalent* if they can consistently be treated in one context as identical, and in another as distinct. This concept has proved to have great power in mathematical argument, since, if it is appropriately applied, it makes it possible to reach higher levels of generality by deliberately ignoring distinctions which, under conditions of greater particularity, are treated as essential; it also has a place in science, enabling words like 'mammal', 'molecule', and 'metal' to be defined to the satisfaction of those concerned, notwithstanding the extreme diversity of their exponents. In general, however, equivalents can be distinguished on closer scrutiny, even if in mathematical contexts this is axiomatically forbidden, or in observational situations technically impossible.

Equivalence is therefore a technique and not a theory. That is, the distinguishing fact, that the parity-relation between two objects, $=$ or $\neq$, may depend on the context in which they are considered, is interpreted as opening two alternative theories of such objects, either of which is self-consistent, but which cannot be coasserted without contradiction. It is a method of replacing one theory by another, according to the end in view, rather than a single theory applicable in cases where the 'end in view' is itself an effective component of the material in hand. Physicists are content to say, given a bubble-chamber record, that "one + pion made this track, and a slower one made the other", treating the two pions just like two bullets, while in their theoretical moods this kind of distinction is forgotten and every + pion is made to appear identical to every other.

The question I am here concerned with is this: Can one explain the observable distinctions among things as in any part due to the manner in which we perceive a certain class of things which are *fundamentally* indistinguishable? If so, the latter might be usefully hypothesized as an underlying level of existence, which might help to explain some of the phenomena observed at more accessible levels. It is clear that this question can't be treated except within a single self-consistent theoretical framework. If the supposed 'indistinguishables' were taken as merely 'equivalents', we would be entering the underlying level only by throwing away information, whereas I want to see whether information could emerge somehow in the process of ascending from the lower level of indistinguishables, to the higher one of distinct entities.

This would be a vain hope if the concept of equivalence were adequate, for it would amount to inferring particulars from generalities. I shall show that there is an alternative, and that many things can in fact be inferred from it; I cannot *prove* that these inferences explain anything in the physical world, for that is not a mathematical matter, but one of interpretation; but I shall certainly make some of it look extremely plausible.

This alternative is the concept of 'indistinguishability'. This involves introducing a third parity-relation, besides identity and distinction, which may hold between mathematical entities; I propose to call any theory of this kind *triparitous* in contradistinction to the familiar *biparitous* theories. I shall denote the three parity-relations by the following symbols:

$= ab$ means that 'a and b are identical'

$\overset{\wedge}{-} ab$ means that 'a and b are *twins*'

$\overset{\shortmid}{-} ab$ means that 'a and b are distinct'.

Since there are three of them, their negations are unions of the other two, making six parity-relation altogether; thus:

$\neq ab$ means that 'a and b are non-identical'

$\hat{\dotplus} ab$ means that 'a and b are *bipar*'

$\overset{|}{+} ab$ means that 'a and b are indistinct'.

If $\overset{\wedge}{-} ab$, i.e. if a, b are twins, the entities they denote are said to be indistinguishable. The peculiarity of indistinguishables is that as members of any one class they contribute severally to its cardinality (i.e. they can be counted, though not by the same methods that work for say eggs), whereas in any other situation they behave in the same way as identicals would.

1.B. PRIMARY AND SECONDARY INDISTINGUISHABLES

It will be shown in Chapter III that any kind of observation delivers information in a biparitous form; it is not possible therefore to rely directly on observation to distinguish between equivalents and indistinguishables. We shall therefore have to formulate some kind of hypothesis having testable consequences if we are to make use of the concept of indistinguishability, and, as a necessary preliminary, to determine more precisely the competence of observation in this regard.

As I have mentioned, equivalents can in general be distinguished by closer empirical scrutiny. These then are potentially recognisable as such by observation. But this is not a logically necessary condition, and there could be equivalents which no scrutiny could distinguish.

Such equivalents I shall describe for the nonce as 'inscrutable'. Indeed, when electrons are said to be equivalent, they are also assumed to be inscrutably so; nothing, apart from accidents of place or behaviour, distinguishes one from another. This class of entities, seemingly not an empty one, is then subject to confusion with indistinguishables in any observation; on the other hand, inscrutables can (theoretically: not necessarily technically) be distinguished from formal equivalents of the kind whose identification requires potentially available information to be ignored.

It appears to be the case empirically that all instances of credibly inscrutable equivalence are known only by observation of events. On the one hand, entities which are observable without seriously altering the system observed, that is, entities on any but the atomic scale or below, seem always to reveal differences of some kind on sufficiently close examination. Events also, even at the subatomic scale, have this character, differing in energy, angle of incidence of particles, and the like. On the other hand, subatomic particles – the presenting case of inscrutability – are generally recognised to exist only as inferences from observation of the events in which they participate. They are both too small to suffer from 'imperfections' and too hypothetical to exhibit the 'accidents' (in the older sense) which can distinguish events. Briefly, we may say that particles hypostasize causal relations between events; logically, they are secondary constructs; and may therefore be inscrutably equivalent.

Fortunately, the kind of 'events' which are involved in this relationship with particles seems sufficiently recognisable in practice, mainly by the small scale of the observation and the high energy-densities usually encountered; they are in no danger of being confused with events on a more familiar scale. We may then further postulate that all particles participating in events of this kind (the only events I shall be concerned with) are members of inscrutably equivalent classes. We have therefore to distinguish between three kinds of classes of entities: (a) distinguishables, (b) inscrutables, and (c) indistinguishables, not observed or inferred from events and *a fortiori* strictly unobservable, though not necessarily without observable effects.

The theory of indistinguishables is designed to apply primarily to class (c), which I shall therefore call *primary* indistinguishables; but logically it is equally applicable to the class (b), hitherto described as 'winscrutables', notwithstanding that these, unlike primary indistinguishables, could also be taken as merely equivalent. I therefore call the class (b) *secondary* indistinguishables; their main exemplars are the various species of so-called elementary particles.

1.C. CLASSES OF INDISTINGUISHABLES

It will be recalled that I described the peculiarity of indistinguishables in terms of their being non-identical as members of one class, but not (demonstrably) so across a class boundary. This idea can be formulated rather more rigorously as follows:

Let C be the union of classes C′ and C″
Let C′ contain $c_1 \ldots c_n$ together with a member a
Let C″ contain $c_1 \ldots c_n$ together with a member b
Then, if a, b are

the same entity	C, C′, and C″ are identical classes, with cardinal $n + 1$ (1.1.1)
different entities	C, C′, and C″ are three different classes, with cardinals $n + 2$, $n + 1$, and $n + 1$ respectively (1.1.2)
indistinguishable	C, with cardinal $n + 2$ is distinct from C′ and C″ which behave as identical (in the absence of further information) with cardinal $n + 1$ (1.1.3)

Any indistinguishable which might be encountered outside any class can be considered as in a singular class of its own; it is evident from (1.1.3) that such a one can never be proved not to be a member of any class containing any of its twins. But this question would be immediately decided if such a lone indistinguishable were observed along with any other class: for if it were seen to be separate, it would be proved distinct (case 1.1.2), and if not it would not be observed at all. We are thus led to the conclusion that indistinguishables might be 'observed' (i.e. have observable effects) *as classes*, but could never have such effects as individuals.

Since secondary indistinguishables, as defined in 1.B., are observable, in the sense just denied to primary ones, we must infer that if the theory is to be consistent, they must be in themselves describable as classes of primary indistinguishables. Such classes are clearly of a special type, and I shall call them *aggregates.*

This is an important principle. Though argued here with an appearance of rigour, the statement that secondary indistinguishables are aggregates of primary ones is strictly an axiom: for the mathematical theory of indistinguishables cannot contain explicitly any notion of being 'observed' (an extra-mathematical concept), and without this it cannot figure as a theorem. I shall, however, make no use of it till we reach Part II. Nevertheless, the discovery that we have different types of classes of primary indistinguishables prompts a consideration here of what other types there may be, and what criteria are required to distinguish them.

1.D. CORRELATION AND PREDICATION

It is abundantly clear that the existence of indistinguishables is a hypothesis, for which I have so far adduced no evidence. In order to collect such evidence, we shall have a rather elaborate procedure to go through. This is the subject of Chapter III; but at this point it is necessary to point out that whenever physicists postulate the existence of something, they assume that they are establishing a concept, in terms of which they may be able to predicate quantifiable characters of the entity open to experimental verification. Indistinguishables are different in this respect also. First, we shall find that not every postulated class of indistinguishables can be accepted as 'observable' (Chapter III will explain why, Chapter VII will discover which); and secondly, since we want our indistinguishables to explain, if they can, a convincing variety of known facts, our first task on finding an empirically verifiable class will be to try to correlate it with a known physical concept. Since classes of this kind have hardly any properties beyond cardinality, the most important clue here will be the matching of cardinals. This step I call *correlation.*

But it is not the only criterion. For once we have picked on a correlation with some physical concept, we shall come up against a limited class of things which can be said about it. In most cases, these will consist in predicating numerical values in each particular instance and they will be constrained by existing practice in physics; we have therefore to check that such predications will not be inappropriate in the light of any other correlations we may arrive at. For this reason it is expedient to regard *predication* as a separate step from correlation. The classification of types of classes of indistinguishables rests on the cardinality of the proposed correlate and the kind of information appropriate to predicate of it; though not fully perspicuous without further explanation, this classification is tabulated in (1.2).

Most of this material will be made use of later on. The two most important types for the present purpose are the Aggregates, already briefly explained, and those which, because their members are usually called 'degrees of freedom', I designate as 'Liberties'. The latter are employed as indicated in (1.2) for locating a point in space. In the case of a 3-space, for instance, this is done by a vector with three components, each normally a real number giving the distance of the point from each of three reference planes, whose position is understood to be somehow marked or defined in an operationally satisfactory manner. These real numbers are supplied by *predication*: someone, either reporting an observation or framing a hypothesis, assigns them values. Before this is done, we have simply an

Cardinal of correlate	Finite			Infinite	
				Progressive	Simultaneous
Range of predicates acceptable	Binary	Finite	Infinite	Binary	Binary
Type of correlate class	Threshold Limit agg.	Specification Aggregate	Position Liberty	P. O. Set Ordered	Wavefunction Disordered
members	Contributors	Monads	Degrees of freedom	Events	Points
descriptors	Presence 0 or 1	Attributes (e.g. quantum numbers)	Coordinates	Antecedents	Responses
Character of members	Secondary	Concrete	Abstract	Concrete	Abstract
Examples	Smallest possible black hole	Baryon	Location of an event in space	Causal history of an event	Particle moving in a field

(1.2) Table.

unspecified 3-vector; this embodies the idea that three coordinates are required, but no more, to specify any position, once we are given a system of axes to refer them to. This concept, essentially that there are three axes which could be 'marked', comes with the step of correlation of the three axes with the correland class of three indistinguishables. Since the axes are supposed to be marked, they are in an empirical and hence in a biparitous context, whereas the indistinguishables are of course in a triparitous one. Correlation is the step that crosses this barrier; but the results cannot be fully tested without carrying out the second step of predication, which normally requires some form of measurement. As this is a physical operation, my theory cannot throw any light on it, so all these 'tests' of the theory are in a sense incomplete or indirect until supplemented by a truly physical theory.

How this may be done will be examined briefly in Chapter III; the execution lies outside the scope of this work. Meanwhile, the testing of the theory of indistinguishables rests largely on the minimal carry-over of *numbers* to the empirical arena, such as the number 3 in the above example. Of course, such a result as this is of little evidential value; my case rests on the presence of a few much more impressive numerical correlations which will be expounded in due course.

The classification (1.2) applies to both primary and secondary indistinguishables, though the latter occur only in limit aggregates (col. 2) and perhaps ordered classes (col. 5). Secondary indistinguishables can be assigned to a *species* and a *state*. The species is defined by the attributes predicated of its constituent monads, and the state by the coordinates predicated of the degrees of freedom in an appropriate 'liberty'. Primary indistinguishables, having no constituents, are assigned to no species; but they can be regarded as being in some state, provided that we don't postulate any 'change of state' for such entities whose continuity through such a change cannot be defined. Such a 'state' is simply what is predicated of a primary indistinguishable on a given occasion. Continuous existence implies stability, which is a strictly physical attribute and attaches to physical forms, which (primary) indistinguishables are precisely and by definition not.

1.E. THE TRIPARITY OF NOTATIONS

Triparitous mathematics is a wider concept than the theory of indistinguishables to which it is applied here. An example of a triparitous system which is comparatively familiar, though there is little profit in considering it as such, is to be found in ordinary mathematical notation. The three parity-relations can be illustrated by the symbols of a simple algebraic formula. For example, in the expression '$a + \frac{a+b}{c}$' there are three *different* identifiers a, b, c and two *different* operators $+$, ——. The two occurrences of a and of $+$ are *indistinguishable* symbols; the divisors of the second a and of b are *identical* symbols c. Admittedly, we could equally, and more conventionally, describe the two a's and $+$'s as equivalent, but if so they are inscrutably equivalent, for whatever distinguishing marks you may discern in them are by definition the fault of the printer, no more than imperfections in realizing an ideal 'a'.

This excursus into Platonic phraseology is intended to make a point. Idealist philosophy is criticized for not describing the real world, and is rightly discredited; but what if there were some field where it could be used literally? Applied to mathematical notations it is pedantic and unhelpful; but my hypothesis is that in the foundations of the world, a context antipodal to that of moral and aesthetic values where the notion was born, there are things of which the real and ideal coincide. Here one need make no apology for treating inscrutables as indistinguishables; the categories are clear-cut, their properties plain, and observation either impossible or infallible.

Nevertheless, there is one advantage to be gained from using the terminology of triparitous description in relation to notations, namely that we shall have to reject the customary statement that in any one formula 'the same symbol always denotes the same referent', on the grounds that the word 'same' is there used in two senses, which is not the case unless the notation is considered as triparitous.

I shall say instead that 'identical symbols denote identical referents' (failure of which, as the example above shows, would amount to permitting puns in mathematics!), whereas 'indistinguishable symbols denote either identical referents, or twins, according to the context'. This convention may at first seem irksome and unnecessary; if the meaning of the formulae is unambiguous and the theory self-consistent, why should we not use, say, a system of diacritics to distinguish twin referents? Of course, this would be possible; but it has two disadvantages which have persuaded me against it. First, our formulae, already full of new symbols and often rather long, are better spared any avoidable graphical complications. Secondly, the reader would still have to check at each step that the diacritics were rightly placed, and this work is equivalent to deciding where they ought to have been put if they had been left out (though experience with misprints in conventional mathematics falsely suggests it should be much less). As I shall show, we can find contextual rules by which it can be decided in every case which indistinguishable symbols denote identicals, and which may denote twins. Thus the actual presence of diacritics is superfluous, and with experience becomes confusing; it is moreover psychologically easier to doubt inferences, which are invisible, than to question the correctness of complicated formulae which are present to the eye. I could get away with errors of argument more easily if I used diacritics since without them you will have to work out for yourself part of the meaning of each statement.

Linguists, if any are with me, will notice that, with the convention proposed above, my notation will not be 'context-free' (the use of diacritics would disguise the fact but not obviate the principle). This introduces a serious complication into the theory – for it is a feature of all normal mathematics that it is couched in a context-free notation, subject to only occasional rule-bound exceptions, as when we write $\sin^2 x = (\sin x)^2 \neq \sin \sin x$. Loss of this rule imposes an unaccustomed and unwelcome strain on the reader. Even Intuitionist mathematics, which also imposes some mental strain on its devotees, gets by with a context-free notation. It necessitates a whole preliminary section of the work, in which the proper rules for the notation are worked out from first principles – my Chapter II. One can't hope that this will help to popularize the theory.

I.F. LEVELS OF NOTATIONS

In setting out any mathematical argument, we use, if only implicitly, at least two different levels of notation. The *substantive* notation contains expressions denoting relations within the object-theory; for example, in elementary arithmetic, $a < b$ would be such an expression in the substantive notation. But a 'theory' requires that inferences be drawn from statements, and this involves the existence of a different *inferential* notation, statements of which have a different form, typified by $a < b . \Rightarrow . a - b < 0$. The operator $\Rightarrow$ is no part of arithmetic; but it functions indispensably in stating the logical structure of the theory. The inferential notation is not usually developed very far, and may be relegated to the care of discursive prose; it is regarded as the province of logic rather than of mathematics narrowly considered. But its existence cannot be ignored, and it is best when functioning in the clear.

The relata of the inferential notation denote relations in the substantive notation. Denotation is not ostension; but the Laputan custom of confusing them is ingrained in mathematical usage. Thus, the example of a statement in the inferential notation given above uses for its relata graphical copies of relations in the substantive notation. Logic would here require us to prefer 'if X denotes $a < b$ and if Y denotes $a - b < 0$, then $X \Rightarrow Y$'; it is obvious that the Laputans have the edge on this kind of logic, but we should nevertheless include explicitly, as a rule of the inferential notation, that denotation of relata is by graphical copying.

This usually of course 'goes without saying'. But in our case it needs stressing, because we shall need *two* substantive notations, and not one only. Besides the *objective* notation, in which the theory itself is expressed, we shall need a different *semantic* notation, in which to work out the rules for the context-binding, which as we have seen must operate in the former. Its relata will denote symbol-strings (understood as if formulae of the objective notation), and most of its relations will denote semantic equivalence, or the lack of it, between such strings. Unlike the objective notation, it must be context-free (else it would only begin a recursive series of notations) and biparitous (else it could not be context-free). It could, like the inferential notation, use for its relata graphical copies of the symbol-strings denoted; but since these would be copied again at the inferential level, there would be opportunities for confusion unknown in Laputa, which I have found to impede understanding. I therefore enclose graphical copies in square brackets [ ] to create symbols for symbol-strings in the semantic notation.

Henceforward I shall denote the objective notation by $\mathscr{T}$, the semantic notation by $\mathscr{U}$, and the inferential notation by $\mathscr{V}$. It will be possible to keep the first two apart in the text, $\mathscr{U}$ in Chapter II and $\mathscr{T}$ elsewhere. For this reason, it will not be necessary to make the two notations formally distinct. But the same inferential notation will be used alongside both of them, and it is therefore expedient to provide a simple rule by which this one can be distinguished from either of the two substantive notations. This I do by using interposed relators in statements of $\mathscr{V}$ as against preposed relators throughout $\mathscr{T}$ and $\mathscr{U}$. This usage will be especially beneficial in $\mathscr{T}$ as a means of simplifying the syntactic recognition of the relations governing its context-dependence.

1.G. SYNTACTIC SPECIFICATION OF THE NOTATIONS

I shall now set out, for reference, a brief syntactic specification of each of these three notations, incorporating some terms which it will be useful to have defined. The semantic and objective notations follow the same rules, with minor variations as given; but the inferential notation is best set out separately.

Common to both $\mathscr{T}$ and $\mathscr{U}$ are:

N : A *formula* is a string of *symbols* in which an ordered sequence of *arguments* is preceded by a *governor*, where

N1 : A *symbol* is either a character taken from a predetermined inventory accompanied by a definition which includes specification of its *valency*, or a character-string likewise defined; where

N1.1 : The *valency* of a symbol is the number of arguments following it, and

N1.2 : an *identifier* is a symbol of zero valency, whereas

N1.3 : a *functor* is one with non-zero valency; and

N2 : an *argument* is any formula which is a proper substring of another formula, to the definition of whose *governor* it is acceptable in its place; and

N3 : the *governor* of a formula is a symbol whose definition specifies its *domain* and how to determine from its arguments what element of its *range* it denotes; where

N3.1 : the *domain* of a symbol is the direct product of the *ranges* of its arguments, while

N3.1.1 : a *superfunctor* is a functor whose domain includes functors, a formula governed by a superfunctor being a *superfunction*; and

N3.2 : the *range* of a formula is the class of values it may denote; and

N3.2.1: a *relator* is a functor whose range is the class of truth-values, and governs a *relation*; and

N3.2.2: an *auxiliary* is an identifier whose range is empty (such symbols serve as reading-aids, and a formula is deemed unchanged by their omission)

(1.3) Table

Peculiar to $\mathscr{T}$ are the following usages:

T1: the symbols

T1.1: ,) (are auxiliaries of the notation, and

T1.2: $\hat{\_}$ is a bivalent relator denoting twinship;

T2: The null symbol

T2.1: is a relator of the notation, whose

T2.1.1: domain is pairs of relations, and whose
T2.1.2: value is their logical conjunction,

whence it follows, to preserve consistency, that

T2.2: the null formula [which from T1.1 may be written as ( )] denotes 'true', and

T2.2.1: by graphological analogy the symbol (/) denotes 'false'. (1.3a)

Peculiar to $\mathscr{U}$ is the usage that

U1: the symbol [denotes the symbol-string following up to the next], understood as if a formula in $\mathscr{T}$. (1.3b)

As a general mnemonic, the usage in respect of preposed functors which happen to be isographic with their conventional interposed equivalents is that the second argument after the former corresponds to the first argument of the latter, that is to say

R a b is equivalent to b R a (1.3c)

The following specification applies only to $\mathscr{V}$:

V : A *statement* consists of a *protasis* and an *apodosis* separated by an *implicator*, where

V1 : the *protasis* is a relation or sequence of such, copied unchanged from $\mathscr{T}$ or $\mathscr{U}$, or a logical function of protases, in which

V1.1 : every symbol is understood as subject to any condition expressed earlier in the protasis, and

V1.2 : expresses the conjunction of the relations where no other function is indicated; and

V2 : the *apodosis* is a sequence of one or more relations copied from the same notation as used in the protasis, expressing their conjunction; and

V3 : the *implicator* is either

V3.1 : the symbol $\Rightarrow$ which asserts that the apodosis is true if the protasis is true, but otherwise undecidable; or

V3.2 : the symbol $\Leftrightarrow$ which asserts the logical equivalence of the protasis and apodosis.

V4 : no symbol may appear in the apodosis whose range includes twins but is not otherwise specified, unless it is also directly dependent on the governor of the protasis. (1.4)

It should be noted that $\mathscr{V}$ rests on the axioms of classical logic, that is to say, any decidable proposition is either true or false but not both, and there is a bivalent functor $\mid$ over truth-value such that if p, q are any decidable propositions, $\mid\mid\mid pppq$ contradicts q and vice versa. In this logic, the statement $q \Rightarrow p$ asserts that $\mid\mid ppq$ is true; if $\mid pq$ is true, either p or q is false. Thus, if we write p for $\mid pp$, which is true if and only if p is false, we have $=(\ )p \Leftrightarrow =(/)'\,p \Leftrightarrow =(\ )''p$. It is easy to imagine that three parity-relations are somehow equivalent to three truth-values, but this is quite wrong. It might well be that triparitous mathematics could be pursued with the added complication of using say intuitionist logic; but it's bad enough as it is without that. What is not true is false, provided it is decidable at all. However, what we shall find is that *undecidable* relations will arise more frequently, and so more often block off promising starts in the argument than is usual in biparitous mathematics; and this is one reason why the theory that does emerge is noticeably poorer than most mathematical theories (I could say 'simpler', but I refrain.) But not to know whether something is true or false is not the same as to accept that it is neither.

It is also desirable to remember (as always, but usually without explicit remark) that *assertion* is possible only in $\mathscr{V}$. A relation in $\mathscr{T}$ does not assert itself; it has a value true or false, but no locutionary force. To assert for example that a and b are twins, we have the statement in $\mathscr{V}$ that $(\ ) \Rightarrow \hat{\_}ab$, and inferences can be drawn from this; $\hat{\_}ab$ by itself is simply a member of the class of truth-values. If x is a variable, then so is $\hat{\_}ax$, unless we can prove the truth of $=(\ )\hat{\_}ax$, in which case we still have to use the form $\mathrm{P} \Rightarrow =(\ )\hat{\_}ax$ (where P stands for the premises from which the apodosis is established) if we want to formally assert this conclusion.

Statements in $\mathscr{V}$ concerning relations in $\mathscr{U}$ however assert nothing except the putative semantic equivalence or the want of it between two supposed formulae of $\mathscr{T}$. Nothing is conveyed about the truth or even about the well-formedness of the symbol-strings referred to. If this is lost sight of, the arguments in the next chapter may at times seem obscure or perverse. What we now encounter is as near to a pure formalism as one can well get.

CHAPTER II

SEMANTIC THEORY OF THE NOTATION $\mathscr{T}$

The aim of this chapter is to formulate a system of axioms for the theory **T** which I shall elaborate on in subsequent chapters as a description of that level of being which contains the 'indistinguishables' discussed in Chapter I. Wherever possible however I shall refer rather to a more general triparitous theory ***t***, of which **T** will be a special case. To this end I shall require a semantic theory of ***t***, which I shall call **U** and express in the notation $\mathscr{U}$; in view of the generality ascribed to ***t***, this **U** may be called the semantic theory of the notation $\mathscr{T}$ itself.

The axioms of **U** include certain rules for the employment of the functors **in**, ^, and ", which it is assumed will be valid in relation to any object theory ***t***. The relations in **U** will concern the equivalence or otherwise of relata denoting (potentially at least) expressions in ***t***. An exposition of **U** is needed at this point, because until we know the theorems relevant to the semantics of ***t***, which are theorems of **U**, we can't formulate the basic axioms of the theory **T**.

2.A. THE MEANING OF MEANING

The theory **U** is about meaning, specifically for us meanings of expressions of some triparitous theory ***t*** expressed in the notation $\mathscr{T}$. In order to begin the exposition of **U**, we need to know with sufficient clarity what we mean, in this context, by 'meaning'.

Every mathematical relation, like every colloquial sentence, carries a certain quantity of information. The quantity itself, and how to measure it, are not in question however; all we are here concerned with is what is said. Most of the time, we can confine attention to what is meant by saying that two sentences, or two expressions, mean the same – that is, carry the same information. Now in general a statement, mathematical or colloquial, is conveyed by a structured set of lexemes, in which the structure represents the syntax, and the elements structured carry the 'dictionary' information; the latter is a major problem in linguistics, but for the mathematician is simply disposed of by referring to the stated definition of every term. The structure gives more trouble, since an expression is in general *not* fully represented by a simply-ordered string. It is not sufficient to attach an ordinal number to each separately meaningful symbol or group of symbols; but it is certainly sufficient to attach a *finite set* of such ordinals. Such a sufficient set I shall call a *structor*. Every statement in any self-consistent notation can thus be translated reversibly into an *unstructured* set of pairs, each comprising a symbol and a structor, which I shall call *items*.

If care is taken to refer back, from the use of each symbol in any given occurrence, to its original definition, the semantic equivalence between any two expressions can in principle be assessed, asking whether they translate in this way into the same set of items, or at least into mutually transformable sets, once we have defined the idea of a transformation. Thus the meaning of any expression is represented as a set of 'items'; and the semantic theory **U** is an application of set theory to such sets. It is, of course, biparitous: to use a theory which was triparitous here would clearly be circular; it is conventional set-theory we have to use. For the same reason, we have to insist that the notation $\mathscr{U}$ (which we write **U** in) is, like ordinary mathematical notations, context-free. These conditions ought to be stated as an axiom (whose content applies equally, albeit unstated, to almost all mathematics other than triparitous):

> All expressions in **U** are context-free formulae of $\mathscr{U}$ involving exclusively biparitous symbols. (2.1) Axiom

The indefinables of **U** comprise all those of finite set theory, together with the basic concept of 'items' described above; the functor **in** is then defined, as follows:

> If S is a symbol-string in $\boldsymbol{t}$, the formula **in**[S] in **U** denotes the set of items containing the information, if any, conveyed by S in the theory $\boldsymbol{t}$ (2.2) Df. **in**

No restriction is placed on the nature of S, but there must be some theory $\boldsymbol{t}$ stated in the context, in whose terms S may (or may not) carry meaning; if S is meaningless in $\boldsymbol{t}$, then = **in** [S]$\emptyset$. For us, we shall usually have **T** in mind as the value of $\boldsymbol{t}$. The use of **in** is subject to the following 'axiom of further information':

> If ST is a symbol-string in $\boldsymbol{t}$ comprising substrings S and T, then
>
> $\leqq$ **in**[ST] $\cup$ **in**[S] **in**[T] (2.3) Axiom

where as before [S] *denotes* the S of t in **U**. Note that the information conveyed by ST can be greater than the union of that in S and T separately, which is evidently the case whenever their ordering is relevant, for this must be represented by an ordering-item not belonging to either **in**[S] or **in**[T]. Any syntactic rule, however, such as N in (1.3), which *prescribes* the order of symbols, entails that the ordering so prescribed carries no information, and then we have

$$= \mathbf{in}[ST] \cup \mathbf{in}[S]\,\mathbf{in}'[T].$$

2.B. THE PAIR FUNCTORS OF U

It follows from the above remarks that we shall need to have means of distinguishing, in the notation, between pairs of symbols or strings whose ordering is meaningful, and those for which it is not. We can do this by using the following two functors:

There is in **U** a functor ˆ whose domain denotes pairs of symbol-strings [X] [Y], such that

$$= \mathbf{in}\,\hat{}[X][Y] \cup \mathbf{in}[X]\,\mathbf{in}[Y] \qquad (2.4.1)\ \text{Df. }\hat{}$$

There is in **U** a functor ″ whose domain denotes pairs of symbol-strings [X] [Y], such that if $\ddagger$ [X] [Y]

$$\ddagger\,\mathbf{in}''[X][Y] \cup \mathbf{in}[X]\,\mathbf{in}[Y] \qquad (2.4.2)\ \text{Df. }''$$

(The same symbols are also used for functors in **T**, so it is important not to forget the [ ]!).

In Set theory it is more usual to find a definition for ″ in terms of ˆ, but in the present context this relation figures more logically as an axiom, as follows:

For any symbol-strings X, Y in a theory t

$$= \mathbf{in}''[X][Y]\,\mathbf{in}\,\hat{}[X]\,\hat{}[X][Y] \qquad (2.4.3)\ \text{Axiom}$$

It might be supposed that Axiom (2.4.3) was in contradiction with the definitions of ˆ and ″ just given; for if we substitute (2.4.3) into (2.4.2), and apply (2.4.1), we seem to infer that $= \mathbf{in}''[X][Y]\mathbf{in}''[Y][X]$. However, we are here using the symbol t to denote a triparitous theory, and therefore a context-sensitive one, and we cannot assume that X before Y and X after Y will mean the same thing; in the context-free notation of **U** this comes out as a difference information-content between the two [X] in

ˆ[X] ˆ[X][Y]. Indicating this by a suffix $_1$ on the first, we have from (2.4.3) = **in**'' [X][Y] **in** ˆ$[X]_1$ ˆ[X] [Y], which by (2.4.1) = ∪ **in** $[X]_1$ ∪ **in** [X] **in**[Y]; whereas similarly = **in**'' [Y][X] ∪ **in**$[Y]_1$ ∪ **in**[Y] **in**[X]. These are of course equal if = [X][Y], but as this is explicitly repudiated in (2.4.2) there is no contradiction.

However, it might be asked, might not the two strings X and Y have the same meaning in the initial position, and yet not be equivalent elsewhere? This is indeed conceivable; but (2.4.3) forbids it, and therefore requires a convention valid for any triparitous theory ***t*** which (being merely a convention) restricts the range of such theories for which Axiom (2.4.3) can be accepted. It is important to state this explicitly, and to prove that it introduces no inconsistency in our theory **T** in particular. This will be done by showing that the following convention still holds after making free use of (2.4.3) throughout.

> If X is an initial substring of any formula in **T** and if Y can be validly substituted for this occurrence of X, then it can be substituted for any later occurrence of X in the same formula with equal validity (2.5) Axiom

This is asserted only for formulae, since there is no occasion for applying (2.4.3) to any other kind of strings. Its main effect is to authorize the substitution of a single symbol for any substring in a formula which occurs also initially in the formula. It would be shown inconsistent in **T** only if there were an instance where the procedure would alter the interpretation of one or more other symbols in some formula. This does in fact happen for *non-initial* substrings (an example is given in 2.J below), so that the admissibility of the convention is not a foregone conclusion; but I have found no instance where (2.5) as stated has any untoward effects on the development of the theory.

2.C. CLASSIFYING FUNCTORS

The pair functors ˆ, '', also appear in **T**. They are there not identical with their isographs in **U**, and if the **T** functors were to appear in **U** they would be written [ˆ], [''] by convention U1 (1.3b). But in **T** they have the property, which they share with several other functors, of making classes out of their arguments; that is to say, $\in\hat{}\,xy\,x$, $\in\hat{}\,xy\,y$. It is clear enough, from the original characterization of indistinguishables in (1.1), that such functors will have special properties in any triparitous mathematics.

Since **U** has to concern itself with the context-sensitive notation $\mathscr{T}$, in particular with those contexts where a given symbol may denote a referent twin but not identical with that denoted by another indistinguishable symbol; and since one such context is, by the definition (1.1.3), when one of the indistinguishable symbols denotes member of a class to which the other may not belong, it is evident that functors of this type must be noticed in **U** as well as in **T**. I therefore define

> If C is a functor in ***t***, and if *s* is a symbol-string in ***t*** such that C*s* denotes a class and $\in$C*s* *x* is true in ***t*** where *x* is (or is denoted by) some substring of *s*, then C is called a *classifying* functor of ***t***, $\in$**Cl**C (2.6) Df. **Cl**

If therefore B is any bivalent functor in a triparitous theory ***t***, and if $\in$**Cl**C, the values denoted by the two *x*'s in B*x*C*x* may be non-identical twins.

This then is one context, the first we have explicitly encountered, where indistinguishable symbols may not denote identical referents. It also shows up the context-sensitivity inseparable from triparitous systems; for the same formula, in the case where '$\in$**Cl**C, will be proved below to require *identical* referents for the two occurrences of *x* (see (2.12)). Indeed, the main problem in this chapter is to find when indistinguishable symbols do and do not denote identicals.

Note that (2.6), where no [ ] are used, does not apply to **U**; for by Axiom (2.1) this is a biparitous theory, for which the concept has no special significance. But the pair functors ˆ," do have special features, which must now be considered. To use in **U** the form " [*x*] [*y*] presupposes that in ***t*** the symbol strings *x* ... *y* and *y* ... *x* are not generally interchangeable, though both might be used meaningfully; for if only one of the two actually occurs, say *x* ... *y*, then the string *xy* could be equally well and less misleadingly denoted in **U** by ˆ[*x*] [*y*]. But the truth-value of expressions in **U** is independent on what is the case in ***t***. Thus, if it should happen to be true that the string *y* ... *x* does not occur in ***t***, the expression " [*x*] [*y*] is not invalid for this reason; at the worst, it is vacuous (like saying all unicorns have golden hooves). The identifiers in **U** do not necessarily denote well-formed expressions in the object theory, but *any* arbitrary symbol-strings, such only that statements of their equivalence of meaning with formulae of ***t*** may be decidable in **U**. If, however, a particular case of " [*x*] [*y*] is known to be forbidden by a syntax rule of ***t***, it is desirable to signal this fact by writing the pair in the form ˆ[*x*] [*y*], showing that their order is not in question. We can safely identify *xy* and *yx*, if one of the two never occurs.

2.D. INITIAL THEOREMS IN U

Thus far, I have used only expressions in $\mathscr{U}$, mainly axioms and definitions in **U**. I shall now introduce statements involving such expressions, and so begin to need the inferential notation $\mathscr{V}$. The first step towards this is to define the basic relation called a *transformation* of one symbol-string into another. Thus:

> There is in **U** a relator **tf** such that, if [S], [T] are any two identifiers of **U** denoting symbol-strings in ***t***, then

$$\mathbf{tf}[S]\,[T] \Leftrightarrow\; = \mathbf{in}[S]\,\mathbf{in}[T] \qquad (2.7)\ \text{Df. } \mathbf{tf}$$

Remember here that **in**[S], **in**[T] are sets of items, and that equality between sets implies that all and only the members of each are members of the other. It follows that

$$\text{For all} \quad [S], [T] \text{ in } \mathbf{U},\ \mathbf{tf}[S]\,[T] \Leftrightarrow \mathbf{tf}[T]\,[S] \qquad (2.8)\ \text{Th.}$$

Moreover, we have from (2.7) and Axiom (2.1), the result that

$$\text{For all A, B, S, T in } \mathbf{U},\ \mathbf{tf}\ \text{ST} \Leftrightarrow \mathbf{tf}\ \text{"A" SB "A" TB} \qquad (2.9)\ \text{Th.}$$

A key question for the development of the theory is whether a formula containing two occurrences of a given symbol-string can be transformed into one in which this string appears once only. We can prove the theorem that:

$$\leqq S[X],\ \leqq T''[X]_1\,[X]_2,\ = [X]\,[X]_1,\ = [X]\,[X]_2,$$
$$\mathbf{tf}\,ST \Rightarrow\; = \mathbf{in}[X]_1\,\mathbf{in}[X]_2 \qquad (2.10)\ \text{Th.}$$

(Note that the second term in the protasis does not imply that the two X strings are contiguous in T.) Let us write S = [AXC] and T = [DXEXF]; and let us further define B so that **tf**[BX]∅. Then the expression **tf**[S] [T] becomes **tf**[AXBXC] [DXEXF], or more explicitly **tf**[∅AXBXC∅]. [∅DXEXF∅]. It now follows from Theorem (2.9) that **tf**[ABC] [DEF]. If now $\ne$ **in**$[X]_1$ **in**$[X]_2$, it follows from (2.7) that **tf** ST is false, for this implies that = **in**[AXC] **in**[DXEXF] whence from the transformation just proved = **in**[X] **in**$[X]_1\,[X]_2$ whence $\ne$ **in**[X] **in**$[X]_2$ contrary to the protasis of (2.10), so that by elimination we have only = **in**$[X]_1$ **in**$[X]_2$, Q.E.D.

Another useful though bizarre theorem, which while in general meaningless in ***t*** can be proved true in **U**, is as follows:

> If [U] in **U** denotes a functor U of ***t***, such that '∈**Cl** U, we can define another functor V in ***t*** such that, for any x,
>
> **tf** "x[U] ˆ[V]x (2.11) Th.

From Axiom (2.4.3) and (2.4.1), we have **tf** "x[U] ˆx ˆx[U] = ˆˆ[U]xx. From Theorem (2.10) therefore the two x's in ˆˆ[U]xx denote identicals in ***t***. ˆ[U]x does not denote a class containing the referent of x since [U] is not classifying, so we may now write **tf** ˆˆ[U]xx ˆ[U]" xx. But ˆ[U]" xx is the direct transcription into **U** of a Uyy in ***t***, where =[y]x, both y's being arguments under U, and we can define V in ***t*** by = Vy Uyy. Restating all this in 𝒰, we arrive at **tf** ˆ[U]" xx ˆ[V]x, and so by collation we have **tf**" x[U] ˆ[V]x, Q.E.D.

It is important to note that if U is classifying in ***t***, the two x's generated by the first generation, since from (2.4) they carry in general different information, may denote not identical yy but different yz in ***t*** and the transformation to ˆ[U]" xx is not necessarily possible (for = ˆxx" xx if and only if the x's denote identicals in ***t***). This theorem is used in proving Theorem (2.12).

2.E. INDISTINGUISHABLE ARGUMENTS – CASES (a) & (b)

I turn now to consider the main topic of this chapter, namely the identification of the syntactic contexts in a triparitous theory ***t*** where indistinguishable symbols must denote identical referents, as against those where they may denote twins. I shall base this work on a synopsis of the various types of formulae possible in 𝒯, which may be determined from the rules N of (1.3). If B is the governor, then it may be either

that (a) B is the only functor present,
or (b) there are one or more non-classifying functors ≠ B,
or (c) B occurs again, dependent on the governing B,
or (d) B occurs again dependent on another non-classifying functor,
or (e) there is a classifying functor dependent on B,
or (f) B is classifying, and occurs again,
or (g) B occurs again dependent on another classifying functor.

Apart from a few complications, discussed in 2.H, all relevant cases fall under one of the above heads, so that it is (till then) sufficient to dispose of each of the cases (a) – (g).

As always, of course, we consider the various formulae as objects in **U** with no regard to their meaning in ***t***. The strategy of the search will be to try to reduce a general formula having all or most of its arguments represented by *x*, to a form with fewer occurrences of *x*, using successive transformations as described in 2.D. Occurrences which can be merged in this process must carry identical information, as shown in Theorem (2.10). Those ones which resist such merger need not denote identicals, but may refer to twin referents, though we can't prove that identical referents are excluded, which must depend on particulars not in consideration here. What we here discover will be general constraints on what formulae in a triparitous theory ***t*** may mean, arising from the logic of the notation $\mathscr{T}$ which is claimed to be sufficiently general to express any triparitous theory. Since $\mathscr{T}$ is the only notation yet devised for this purpose this claim cannot be proved; though it could be disproved by a counter-example.

I begin with the case (b), from which (a) follows as a corollary. I shall prove that

> If B is a bivalent, and U a univalent non-classifying functor in ***t***, then in the formula B*x*U*x* the two *x*'s denote identical referents
>
> (2.12) Th.

From the rules N in (1.3), the only ordering which is not syntactically determined is that between the two arguments under B, so that by (2.4) we have **tf** [B*x* U*x*] ˆ[B] ''[*x*] ˆ[U*x*] (note that since B*x*U*x* is a formula in ***t***, [B*x*U*x*] is an identifier in **U**, conformably to the definition (2.7) of **tf**). Next, by Axiom (2.4.3) we have further that **tf** ˆ[B] ''[*x*] ˆ[U*x*] ˆ[B]ˆ [*x*] ˆ[*x*] ˆ[U*x*], and by the same Axiom again, **tf** ˆ[B] ˆ[*x*] ˆ[*x*] ˆ[U*x*]ˆ[B] ˆ[*x*] ''[*x*] [U]. Since U is non-classifying, we can now apply Theorem (2.11), and define a new functor V such that **tf** ˆ[B] ˆ[*x*] ''[*x*] [U] ˆ[B] ˆ[*x*] ˆ[V] [*x*] which transforms by (2.4.3) into ˆ[B] ''[*x*] [V]. Now, since V depends only on U and the notational syntax, V too is non-classifying, so we can repeat the last two steps introducing a new functor W such that **tf** ˆ[B] ''[*x*] [V] ˆ[B] ˆ[W] [*x*]. This last is the transcription in $\mathscr{U}$ of a BW*x* in $\mathscr{T}$. We have thus shown that we can reduce B*x*U*x* to an equivalent form BW*x* in ***t***, so that from Theorem (2.10) the two *x*'s in B*x*U*x* convey identical information, Q.E.D.

By substituting I for U, where I denotes the identity functor, the same conclusion arises for Bxx = BxIx as a special case; and likewise by substituting Tx, where T is a trivalent functor, for B, so we may state:

> In the formula Bxx of $\boldsymbol{t}$, the two x's, and in Txxx the three x's, denote identicals. (2.12a) Th.

The case BUxUx is not much harder; if we write A for the string BU, we get AxUx which conforms to Theorem (2.12), and equally since the U's are identical from (2.12a) we have **tf**[BUxUx] [Axx], and this Axx is conformable to (2.12a). Therefore

> In the formula BUxUx of $\boldsymbol{t}$, the two U's and the two x's both denote pairs of identicals (2.13) Th.

This completes case (a), but case (b) has a further type, namely BUxVx with two non-identical univalent functors U, V. I shall prove

> If B is a bivalent and U, V nonidentical univalent non-classifying functors in $\boldsymbol{t}$, then in the formula BUxVx the two x's may denote non-identical values (2.14) Th.

Given the formula, we can define a nonce-symbol / such that **tf**[BUxVx] [BUV/xx], for we thereby delete nothing, and either side can be reconstructed by rule from the other. Since BUVxx would be read in $\mathscr{T}$ as B(UVx)x which is untranslatable into BUxVx, we have $\nmid$ **in**[BUVxx] **in**[BUxVx], whence $\nmid$ **in**[BUVxx] **in**[BUV/xx] by definition, and from Theorem (2.9) $\nmid \emptyset$ **in**[/]. Thus / contains the information derived from the ordering of the symbols. However, we can now analyse the formula BUV/xx into a compound functor, or superfunction, BUV, and a single compound argument written /xx. In the latter, / is construed as a bivalent functor, so that from Theorem (2.12) the two x's here denote identicals, so that /xx carries only two items of information, both non-null; likewise, BUV carries three items, since $\nmid$ UV. Thus BUV/xx has five items, and so therefore does its equivalent BUxVx, Q.E.D.

An immediate corollary is now that

> In the formula TUxVxWx of $\boldsymbol{t}$, the three x's may all denote nonidentical values (2.14a) Th.

2.F. ENCHAINED FUNCTORS – CASES (c) & (d)

Here again I start with case (d), and treat case (c) as a special case. I shall prove that

> If B is a bivalent and U a univalent functor both non-classifying in **t**, then in the formula BB*xx*UB*xx* the first pair of *x*'s denote one and the second pair a possibly different referent; while all three B's denote identical functors (2.15) Th.

First, each of the subformulae B*xx* can be rewritten, from Theorem (2.12), as V*x*, giving **tf** [BB*xx*UB*xx*] [BV*x*UV*x*]; the latter form is conformable to Theorem (2.14), Q.E.D.1. Analogously to the proof of (2.14), we can alternatively recast BB*xx*UB*xx* in the form BBUB/*xx*, *xx* containing the superfunction BBUB which can be successively transformed as follows: from [BBUB] to ˆ[B] ˆ[B] ˆ[UB] to ˆ[B] ˆ[B] ˆ[B] ˆ[U] [B] to ˆ[B] ˆ[B] "[B] [U] to ˆ[B] ˆ[B] ˆ[V] [B] to ˆ[B] "[B] [V] to ˆ[B] "[W] [B] to "[B] [W] to ˆ[X] [B] which transcribes XB in **t**, where the X has the role of a superfunctor carrying all the non-B information from BBUB; as B appears at last once only, there is only one item. Q.E.D.2

The next type of formula also belongs to case (d), and the two following ones to the simpler case (c); none are mere corollaries.

> If A, B are two bivalent non-classifying functors the formula B*x*A*x*B*xx* in **t** requires one value for both B's and another for all four *x*'s (2.16) Th.

B*x*A*x*B*xx* can be reduced by steps each already justified to B*x*UB*xx* or to B*x*A*x*V*x*, and thence to B*x*W*x* which conforms to Theorem (2.12), Q.E.D.

> In the formula BB*xx*U*x* of **t** both B's denote the same functor, and the first two *x*'s denote the same referent not necessarily that of the 3rd (2.17) Th.

The identity of the B's is a corollary of (2.15); writing V*x* for B*xx* gives BV*x*U*x* which conforms to Theorem (2.14), Q.E.D. Making the last substitution twice over, and applying Theorem (2.13), gives a proof of

> In the formula BB*xx*B*xx* of **t** all three B's and all four *x*'s denote the same referents (2.18) Th.

2.G. CLASSIFYING FUNCTORS – CASES (e) to (g)

The simplest type of case (e) is B*x*V*x* where V is classifying. It has already been shown, in Theorem (2.6a), that the two *x*'s here may denote non-identicals. In the type B$VxVx$, each Vx denotes a class containing x, and if x has twins, (1.1) shows that the identity of two such classes is always in doubt. The heuristic, seemingly suggested by (1.1.1), that the union of the two classes has the same cardinality if and only if they are identical, presupposes that they are objects capable of being continuously observed or reliably copied; this proposition is shown later, in 3.C, to be untenable. It is not therefore possible to treat $BVxVx$ like $BUxUx$ where U is non-classifying, and we conclude that

> In the formula $BVxVx$ of $\boldsymbol{t}$, where V is a classifying functor, the two V's denote identicals, but the two x's need not do so (2.19) Th.

Case (f) covers formulae containing $CxCxx$. Since the proof of Theorem (2.12a) does not assume B non-classifying, the two x's in Cxx denote identicals, so we have **tf**[Cxx] [Dx], where D is classifying whereupon $CxDx$ emerges as a special case of Theorem (2.6a) with non-identical x's. The two C's, like the B's in $BxBxx$ where the classifying property is irrelevant, must denote identicals. Thus

> If C is a bivalent classifying functor in $\boldsymbol{t}$, in the formula $CxCxx$ the two C's and the last pair of x's denote identicals, but the first x need not (2.20) Th.

A straightforward corollary is that

> In $CxCxCxx$, all three C's denote identicals, while of the four x's only the last pair need do so (2.21) Th.

The type $CCxxCxx$, also covered by case (f), reduces to $CDxDx$ and falls under Theorem (2.19). Finally, if a non-classifying functor intervenes, we have

> In $CxBxCxx$ where B is non-classifying, the two C's, the first two x's, and the second two x's denote identicals, but the former and the later x's need not do so (2.22) Th.

Turning now to case (g), where we have two different classifying functors enchained as in $CxDxCxx$, the two C's, since one is governed by the classifying D while the other is not, may denote twins; in other respects this formula follows the pattern of Theorem (2.21), so that

> If C, D are bivalent classifying functors in $\boldsymbol{t}$ then in the formula

C*x*D*x*C*xx* the two C's may denote twin functors, and of the *x*'s only the last two need denote identicals (2.23) Th.

This theorem concludes our survey of the basic formulae. All more complex ones (and of course some of those listed above) can be reduced to simpler types by successive steps, following the pattern exemplified in the proofs above.

It is perhaps pertinent to point out that the theorems involving enchained functors, though they can and do apply to real cases, are not wholly realistic, in that only exceptionally are *functors* defined so as to admit of twin values. In this work, the only case among bivalent functors is that denoted by G in Chapter V, which can have two twin values; it does occur in formulae of type (g), but as it happens only in a context where twins do not have to be distinguished. There is thus no genuine case where these theorems matter to us; but, of course, this need not apply to other applications of the theory.

2.H. DECLASSIFYING & CONFOUNDING FUNCTORS

It will be obvious from the foregoing that the interpretation of any formula depends in general on what functors it contains, and especially on whether these are classifying. Before we can properly go on to reduce the results to a syntactic rule, we must consider one or two cases where functors can be simply ignored (for the purpose of determining what values a given symbol may have) or at least lose their classifying property.

First, there are declassifying functors, whose property is that a classifying functor directly governed by one loses its special status. An obvious trivial example is ∈; if C*x* denotes a class containing *x*, and so rates as classifying, ∈C*x* is true of any member of this class so that according to (2.6) '∈ **Cl** [∈C]. A non-trivial example is the Ç defined in Chapter VII, where Çˆ is shown to be non-classifying, though ˆ alone is so. The main problem with such functors is a possible ambiguity of interpretation, arising from the separability of a classifying functor from a non-classifying compound. Thus, it is obvious that if F is declassifying and C is classifying, the two *x*'s in B*x*FC*x* denote identicals. But what about BFC*x*FC*x* ? Does it behave like BU*x*U*x* in Theorem (2.13), or like BV*x*V*x* in Theorem (2.19) ? If we write A for BF AC*x*FC*x* allows non-identical *x*'s by Theorem (2.14); if, as (2.13) warrants, we write Z for BFF, then AC*x*C*x* also allows non-identical *x*'s by Theorem (2.19). Evidently therefore declassification can operate only once among the arguments of a given functor; that is to say:

> If B is a bivalent and C a classifying functor in $\boldsymbol{t}$, but FC non-classifying, in the formula BFCxFCx the two x's nevertheless may denote twin referents (2.24) Th.

Of course, with different de-classifiers F, G, we cannot even guarantee that the C's denote identical functors in BFCxGCx, from Theorem (2.14).

Another case, somewhat analogous to this, arises in definitions. These, as we have seen, are statements in which the definiend occurs for the first time in the apodosis, normally governed by a relator, usually = ; in fact, all definitions can be expressed in this form. In this situation, if the definiend is a functor it is treated as transparent, in that its arguments are regarded as being directly dependent (or 'immediately governed') by the relator. Logically, this is because it is in effect ' = F' rather than 'F' which is the subject of the statement, from which the general usage of F is to be inferred by the usual syntactic rules.

There is also a small class of *confounding* functors, which are so to speak transparent by nature. The definition is that a functor F is confounding if, for any contextually relevant argument x, and any predicate P, P F$x \Rightarrow$ Px ; that is to say, x is always an acceptable value (normally of course not the only one) of Fx. Once again, in reckoning dependence or 'immediate governance', any confounding functor is simply ignored. Again, the reason is obvious: any statement involving Fx can be regarded as a predication PFx, and since this implies Px we are debarred from denying the consequences of the latter form, if the former is accepted as true. The only confounding functors appearing in this theory are $\hat{\circ}$ (defined in (5.2)), its derivative $\boldsymbol{k}$ (5.2a), and the conditional quantification functor defined in (4.12).

2.J. CONCURRENCE OF SYMBOLS

We are now in a position to formulate more systematically the conditions under which indistinguishable symbols (or symbol-strings) defined in a triparitous theory $\boldsymbol{t}$ are constrained to denote identical elements of $\boldsymbol{t}$. What we need to do is to restate the results arrived at in terms of the notation rules N of (1.3) applying to $\mathscr{T}$. This leads to the following table (supporting theorems indicated on the right):

Indistinguishable symbols denote identical referents *either*

in a governor-dependent relationship to each other	
whether directly	15, 17, 18, 20, 21
or indirectly	15, 16, 22
unless a classifying functor intervenes	23; *or*
if both directly depend on one governor	12a, 20
or on indistinguishable symbols which denote identical non-classifying governors	13; *or*
if one depends directly and the other indirectly, with *no* intervening classifying functor, on one governor	6a, 12,
but not if both are indirectly dependent on one governor through non-identicals	14
nor if they depend on identical classifying governors	19

The syntactic relation which we derive from these results can now be formally defined. We may say that indistinguishable symbols *must* denote identicals if and only if they are *concurrent*, where

C: in any formula in $\mathscr{T}$, two symbols s, t are mutually concurrent (written $\| \, s\, t$) if *either*

C1: s is dependent on t, meaning that it is an argument under t or the governor of an argument of

C1.1: the functor t, *or* of
C1.2: one or more non-classifying dependents on t; *or*

C2: s is dependent on

C2.1: the immediate governor of t, *and not on*
C2.2: any classifying functor dependent on this functor; *or* if

C3: s and t are each dependent on indistinguishable concurrent non-classifying functors with no intermediates (2.25) Df. $\|$

The relation of concurrence is thus defined for any two symbols, not for indistinguishable ones only. The rules C above define the context-dependence characteristic of $\mathscr{T}$ and all instances of it can be traced to this cause. Examples illustrating the use of the rules are given in 4.B, along with a method of checking them.

The effects of concurrence on the denotation of symbols can be summarized as under:

Indistinguishable symbols	
if concurrent, denote	*identical* referents
if non-concurrent, ought only to be used for	*indistinct* referents
Different symbols	
if concurrent, normally denote	*non-identical* referents
if non-concurrent, give no information	(2.26)

The relation $\|$ of concurrence is transitive over sets of indistinguishable symbols, but not over distinct ones. Thus, in the formula $\mathrm{BU}x\mathrm{V}y$, all four of B, U, x, V are concurrent, as are all of B, U, V, y; but x, y are not concurrent with each other. On the other hand, if T is a trivalent functor, in $\mathrm{T}x\mathrm{U}y\mathrm{V}y$ each of the y's is concurrent with the x, but not with each other. For this reason we need a new term to designate the relationship among the symbols in such a formula as this. This can be done as follows:

> Two symbols whose domains are identical and whose ranges are not disjunct are called *isotypic* (2.27.1)

> A *nexus* is any set of isotypic symbols in one formula which are connected by concurrence (2.27.2)

Thus, the x, y, y in $\mathrm{T}x\mathrm{U}y\mathrm{V}y$ form a nexus; the x, y in $\mathrm{BU}x\mathrm{V}y$ do not because they are not concurrent. But in both cases it is likely that the functors will have overlapping ranges, and if so U, V will constitute a nexus in both; but B, T having a different valency and therefore different domains will not belong to the nexus.

2.K. CONCURRENCE IN $\mathscr{V}$

The syntactic unit with which we shall most often be confronted is the statement in $\mathscr{V}$, and we have therefore to consider how the concurrence relation operates here. This is by no means obvious from what has gone before; but it is possible to reduce such statements to a form to which the results set out in 2.J can be applied.

A statement in $\mathscr{V}$ asserts that the protasis **P** is a sufficient condition for the apodosis A to be true; this may be symbolized as **SPA**, where **S** is a relator denoting 'sufficient condition'. Such a condition is the conjunction of one or more necessary conditions; that is, we may write

$= \mathrm{SP}\,\&\mathrm{NP}_1\,\mathrm{NP}_2\ldots$ where P_1 &c. are separate terms of P, and **N** denotes 'necessary condition'. I shall develop an *ad hoc* formalism in $\mathscr{T}$ for expressing such relations; this is of course no part of **T**, and its use will soon be over.

Let R be a bivalent relator. And let * be a bivalent superfunctor whose arguments are first a relator and second a truth-value, such that $= {*}\mathrm{R}(\ )\,\mathrm{R}$, or, with the arguments y,z included $= {*}\mathrm{R}()yz\,\mathrm{R}yz$. I now define $*\mathrm{R}x$, where x is a truth-value, by the condition $= {*}\mathrm{R}xyz \cup {*}\mathrm{R}()yz = (/)x$; that is to say, if $*\mathrm{R}xyz$ is true, either $\mathrm{R}yz$ is true or x is false, so that $\mathrm{R}yz$ is a necessary condition for x. Any instance of **NPA** can thus be expressed in the form $*\mathrm{RA}p$, where p stands for the arguments of P, whose governor is subsumed in R. Thus we assume $= \mathrm{P}\ \mathrm{R}yz, = yz\,p$, and it is clear that the arguments y, z occur directly dependent on the governor *R and so are concurrent with their indistinguishables arising from the expansion of A.

This holds severally for each term in the protasis, but C2.1 of (2.25) will not be satisfied, unless the whole protasis can be expressed with a single governor – and the mere conjunction of other governors will not serve. This is the reason for the rule V2.1 of (1.4) in the syntactic specification of $\mathscr{V}$. If this can be done, we shall have a form such as $\mathrm{G}yz$, where G is the general governor and y, z represent all its immediate arguments. With a superfunctor $^+$ analogous to * we can define $= {}^+\mathrm{G}xyz \cup {}^+\mathrm{G}(/)yz = ()x$ where ${}^+\mathrm{G}(/)yz$ denies $\mathrm{G}yz$ and ${}^+\mathrm{G}xyz$ says that $\mathrm{G}yz$ is a sufficient condition for x. Thus **SPA** can be written ${}^+\mathrm{GA}p, = pyz$ &c., and as before if y, z reappear in the apodosis they will be concurrent with their indistinguishables in direct dependence on $^+$G in the protasis. Provided of course that no classifying functor intervenes. If no such general governor G can be found, the statement is not vacuous, but some of the symbols in the apodosis will only be defined to within twinship, which may not be sufficient for the argument to progress.

Unfortunately, protases in the 'canonical form' specified by V2.1 are apt to be impenetrably obscure; I have generally given both the canonical form and the equivalent term-by-term format when the latter helps the reader, in subsequent chapters. Since, strictly, one ought not to claim to have understood a chain of argument without having, among other things, checked the concurrence of all relevant symbols through each statement in $\mathscr{V}$, one equally ought not to accept the equivalence between a canonical and an analytic form of its protasis without checking that too. Triparitous mathematics is hard work – it's fortunate that apparently there isn't very much of it!

2.L. COMPOUND STATEMENTS AND QUANTIFICATION

The rule V2 forbids the apodosis of any statement in $\mathscr{V}$ to contain the relator $\Rightarrow$ unless this is defined in **T**. But the simplest acceptable definition of say $Px \Rightarrow Qx$ as a relation in **T** would be to equate it with $'\& \ Px\ 'Qx$ (where as before $'$ stands for 'not'); but in this expression the two x's are non-concurrent, so that a weaker relation than $Px \Rightarrow Qx$ is in fact conveyed; nor can any twisting preserve concurrence. Thus implication, in the sense it has in $\mathscr{V}$, is not strictly definable in **T**, and consequently statements of the form $A \Rightarrow (B \Rightarrow C)$, freely used in biparitous arguments, cannot be expressed where indistinguishables are involved. This restriction is irksome, but far from stultifying; its main impact is on the matter of quantification.

It is obvious that *existential* quantification should present problems in a triparitous theory. To assert the existence of something, as is classically done in the statement $(\forall n)(\exists x)(x > n)$, is to assert that the class of numbers greater than a certain n is not empty, which presupposes the feasibility of testing the assertion, say by examining the class of integers one by one till a bigger one than n turned up – or, as it might be, not. This is understood to be practicable, notwithstanding that the class is infinite, because it can be ordered, so that one knows of each number whether it has been tested or not. For finite n the task is finite in consequence; for infinite n we have to use transfinite induction, which is – perhaps – another matter. But in the case where a class of indistinguishables is in question, this will not work, because indistinguishables cannot be labelled so as to know which have been tested and which not, so the task is not necessarily finite; nor can we have recourse to transfinite induction, if only because the want of labels aborts the Axiom of Choice on which this procedure depends. Therefore, existential quantification is in general impossible in a triparitous theory, except in special cases.

Universal quantification, or something acceptably similar, is however possible. This happens in two ways. First, every term in the protasis of a statement of $\mathscr{V}$ is assumed to be universally quantified (subject to any preceding terms); this is unobjectionable because $\mathscr{V}$ is in itself biparitous, and if it were not so assumed the statements it makes would in general lack implicatory force. Second, it is possible to define a functor in **T** which confers a restricted kind of quantification, which I call *conditional*, on one of its arguments. The definition runs as follows in analytic form (the canonical form (appears below as (2.28a)):

$$\in \mathbf{C}x, \in \mathbf{P} \hat{\cup} \mathrm{PQ}, \in \mathbf{2}\mathrm{Q}x, \nmid {}'\mathrm{Q}x = (\,)\mathrm{P}x \Rightarrow \mathrm{P}, x(\mathrm{Q}x) \qquad (2.28)\ \mathrm{Df.},$$

The first term states that x is a class, which is in this context merely to say that it exists; next, we say that both P and Q are predicates and third that Qx belongs to the class **2** whose members are 'true' and 'false', implying that Qx is decidable. The fourth term says that $'$Qx and $=(\,)$Px are mutually contradictory, so that Qx is compatible with $=(\,)$Px (i.e. with Px being true), while $'$Qx is compatible only with $\nmid(\,)$Px (i.e. with Px being false *or undecidable*). All this can then be written as P ,x(Qx) meaning that P is true of any x such that Qx, but not necessarily for any other x.

(2.28) can be restated in the notation of 2.J to check the concurrence of the various symbols x. I use [ ] here as extra auxiliaries to make plain the structure of the statement:

$$*\in[*\in[*\in[*\nmid[\mathrm{P}, x(\mathrm{Q}x)]\,'\mathrm{Q}x = (\,)\mathrm{P}x]\mathbf{2}\mathrm{Q}x]\mathbf{P}\hat{\cup}\mathrm{PQ}]\mathbf{C}x] \qquad (2.28a)$$

The heart of the statement, the formula P, x(Qx), is embedded four deep in necessary conditions, the first being $\nmid{}'$ Q$x = (\,)$Px and the others being successively earlier terms from the protasis of (2.28). Since x is directly dependent on the initial $*\in$, and since no classifying functors occur, it follows that all other x's are concurrent with the first by C2. Similarly each of P and Q (the confounding functor $\hat{\cup}$ being transparent) are direct dependents on the second $*\in$, which guarantees the concurrence of all their later occurrences. The whole statement can therefore be read in the obvious sense (as if there was no question of twins) as an effective definition of the symbol , denoting conditional quantification of x.

It will be noted that in effect ,x(Qx) means "any value of x such that Qx"; this suggests that x is a variable, but it need not be, for the values considered may include the twins of any one of them. It is important to note that , is an ordinary functor of **T**; it does not produce 'bound variables' in the sense used in biparitous set theory. Bourbaki [3] uses an analogous 'functor' which however defines *existential* quantification, which as explained above is not available for us; according to them, in an expression such as $(\forall x)(\mathrm{P}x \Rightarrow \mathrm{Q}x)$, 'the letter x does not occur' which is enough to show why we can't use this way of thinking here. Symbols which 'do not occur' are outside the concurrence rules. The nature of , needs a bit of further illustration.

2.M. COMPARISON OF BIPARITOUS AND TRIPARITOUS QUANTIFICATION

Consider an elementary statement in ordinary arithmetic notation, such as

$$p \geqq 6 \Rightarrow \frac{p}{2} \notin \{p \mid p^2 = 2^n + 1\} \qquad (2.29.1)$$

This states that if p is not less than 6, its half does not belong to the class of p's whose squares are of the form $2^n + 1$. The statement is false, because if $p = 6$, its half 3 has a square 9 which is of this form. What happens if we try to translate (2.29.1) into $\mathscr{T}$? We first need some new symbols; $+$ and $\div$ will be used in preposed position, and $\because$ denotes exponentiation. For 'the class of all p such that ...' we use the classifying functor / in association with the conditional quantifier $_{\shortmid}$ to indicate the class of all values of what follows, that is 'all values of p such that ...'. The result of this is the statement

$$\geqq 6p \Rightarrow{}' \in /_{\shortmid}\, p(= \because p2 + \because 2n\ 1) \div p2 \qquad (2.29.2)$$

The careful reader will notice that, unlike the original which was a false statement, the 'translation' is true; for the class of all values of p not less than 6 whose squares are of the form $2^n + 1$ is empty, and so $p/2$ is certainly not a member of it. The occurrences of p under $\in/$ (where the de-classifying $\in$ cancels the classifying property of / and so saves concurrence) are subject to the conditions on this symbol imposed in the protasis, which is not the case with the 'bound variable' in (2.29.1). To obtain a correct translation from a context-free into a context-sensitive notation we cannot proceed as above on a word-for-word basis; the original meaning is conveyed by the more complex statement

$$\in \mathbf{I} \hat{\cup} pq. \geqq 6p \Rightarrow{}' \in /, q(= \because q2 + \because 2n1) \div p2 \qquad (2.29.3)$$

where the second variable q, constrained along with p explicitly to be an integer **I**, is used. Of course the normal convention would be to use a separate symbol also in (2.29.1); but the form given is nevertheless logically unambiguous, confusing though it is.

The point of this example is to show that $_{\shortmid}$ is indeed a functor, and does not isolate its arguments from conditions stated in associated terms containing concurrent copies of the symbols used. There is no way of creating bound variables in $\mathscr{T}$; the best we can do is to use endemic symbols which don't appear elsewhere with unwanted constraints. Strictly speaking, even 'universal' quantification is impossible, since every statement begins with a specification of the range of the symbols (unless all are locally defined, when a null protasis ( ) may be acceptable), and we have no exact equivalent of '$\forall$'. Thus in (2.29.2) the symbol p, of which it is only stated that it is $\geqq 6$, could be *anything* for which '$\geqq 6p$' can be seen as meaningful, such as a real number (which would make the statement false again, though not in the sense intended by (2.29.1)!). Such statements are only acceptable as items in a series in which general specifications such as $\in \mathbf{I}\, p$ can be assumed to carry over.

2.N. QUANTIFICATION OF DEFINIENDS

The standard form for a statement in $\mathscr{T}$ used as a definition is one in which the definiend occurs once only, in the apodosis, all other symbols used being already defined or locally quantified. In other, non-defining, statements, the interpretation of the apodosis, which must contain no symbols not mentioned in the protasis except previously defined constants, is for this reason unambiguous except through failures of concurrence. But in a definition, the definiend symbol is understood to be *existentially* quantified.

This sole exception to the rule forbidding existential quantification is due to the prescriptive force attaching to a definition. It does not so much assert that something exists, which might not be so, as command that it shall. Such a statement is not open to test (as far as the definiend is concerned), and is thus exempt from the problems attaching to tests of existence in triparitous classes described in 2.K. Of course, it might happen that the definiend could be proved elsewhere not to exist; this would invalidate the definition, showing it to contradict the axioms, but it may have served its purpose in allowing the non-existence theorem to be formulated.

It is in principle always possible to detect the presence of a definiend in a statement functioning as a definition. But it is clearly convenient if this fact be marked; this I do by appending the notation 'Df.' to the statement. Normally the definiend is copied immediately after 'Df.', as for example in (2.28) above; but where this is a verbal expression it is marked in situ [by being *spaced out*].

Since classical logic underlies $\mathscr{T}$ a statement $P \Rightarrow Q$ is equivalent to $'Q \Rightarrow 'P$. In the case of a definition, the negation of an apodosis of the form 'there is a *d* such that Pred. *d*' is clearly 'there is no *d* such that Pred. *d*', equivalent to 'Pred. *d* with *d* universally quantified. This raises no problems. But the negation of the protasis, if this takes the form of a conjunction of terms, will become a disjunction, and this may suffer from ambiguity in a triparitous context. It is therefore expedient that the inversion of a statement should involve only one term of the protasis. This is legitimated by the rule V1.1 of (1.4), which makes the last term detachable so that '$P_1, P_2 \Rightarrow Q$' is equivalent to '$P_1 \Rightarrow (P_2 \Rightarrow Q)$', and so to $P_1 \Rightarrow ({}'Q \Rightarrow {}'P_2)$. Several instances of the use of such partial inversions will be encountered in the sequel.

CHAPTER III

THE PHYSICAL RELEVANCE OF INDISTINGUISHABLES

Before proceeding to develop in detail the mathematical theory of indistinguishables based on the preliminary ideas set out in the preceding chapters, it will be helpful to consider at this point how it is proposed to use the theory to be developed. Briefly, I propose to offer an application of the theory of indistinguishables to fundamental physics. I hope to show in due course that the application is not altogether unsuccessful; but its success is achieved by an unfamiliar route. It is a by no means novel idea, that there is, underlying the phenomena currently observed by physicists, a more primitive level of existence whose properties may be reflected in many of the hitherto inexplicable facts of the physical world. This 'more primitive level' is here postulated to be that which the theory of indistinguishables describes; and the 'facts' which seem to receive an 'explanation' by this hypothesis, though a very mixed lot, include some of the well-known dimensionless ratios, of widely various orders of magnitude, which are correctly evaluated with gratifyingly small departures from their current empirical values. This chapter discusses the underlying philosophy of the method, and results in the formulation of appropriate questions to which the formal development of the theory will then be guided.

3.A. THE CONCEPT OF 'PLANES'

It is a well-known, though not very well-developed, notion, that the world as a whole exhibits various levels of organization, to which the term 'planes' is sometimes attached. On account of the ill-defined and even anti-scientific way in which this term has been used by some, I would have preferred to call them 'levels'; unfortunately, this term has been extensively used in the literature on discrimination systems (see, e.g., 'On the Physical Interpretation and Mathematical Structure of the Combinatorial Hierarchy' by Bastin *et al.* [4]) to refer to the levels of the discrimination hierarchy (which also occur, albeit somewhat disguised, in the present work). I therefore make no further apology for the term 'planes', but proceed at once to explain what I mean by it.

The field of knowledge in which the competence of the scientific method is currently unchallenged comprises two such planes, the physical plane (PP) and the organic plane (OP). The relation between these two is therefore scientifically discussable; and if any other planes are proposed, it must be shown that they are related to these prototypes in a properly analogous manner. To be convincing, the analogy must be close, but to expect identity would imply a similarity among the planes closer than that of the prototypes would warrant.

The relation of the OP to the PP can be briefly stated by saying that the Organic Plane consists of autonomous fluent forms manifested in substance of the Physical Plane. The terms substance, form, manifest, fluent, and autonomous, all call for explanation. Substance and form go together: there is a substance called hydrogen monoxide, among whose forms are a snowflake, a river, a wave, a cloud; there is a more vaguely defined substance called matter, among whose forms, besides those just mentioned, there are also galaxies, stars, planets, rocks. The verb 'manifest' is used to predicate the relation between form and substance; a wave is a form manifested in water, and water can manifest the form of a wave. Any form, in this context, is either 'statant' or 'fluent'. A statant form consists of a given set of molecules or atoms (if it manifests in physical substance), and is held to persist unchanged so long as all of them remain in place. A snowflake is such a statant form; and so are all those listed as forms of matter. A fluent form on the other hand continually replaces the molecules of its substance with other molecules of the same kinds, and the form ceases to exist as soon as this exchange is prevented. Rivers, waves, clouds, are all fluent forms in this sense. Fluents can further be classified as 'exonomous' (owing their form to contingently pre-existing circumstance, as a river owes its form to its bed), 'hyponomous' (having forms computable from the physical properties *of their substance* with the help of, at most, a few contingent parameters, as with waves), or 'autonomous'.

Autonomous fluents have forms determined by some part or property *of themselves as forms*. Living organisms are autonomous, in that their forms persist (notwithstanding progressive changes such as growth and senescence) thanks to retaining the information carried in the DNA of their chromosomes. They are fluents, since they are continually eating and discarding environmental substance, and if this is prevented they die. They are in fact the only known autonomous fluent forms whose substance is of the PP, and therefore they alone constitute the OP, at least on this planet. But they are not the only autonomous fluents of any kind.

For there is also a third plane, called the Human Plane (HP), which consists of such things as culture patterns, languages, special-interest societies, &c., which have the property and means of self-perpetuation as an item in their specification (which makes them autonomous), and which manifest in a continually changing population of OP entities, viz. human beings (which makes them fluents). The analogy here between the PP/OP boundary, and the OP/HP boundary, is obvious; but it is, as was to be expected, less than perfect. For example, whereas the autonomy of OP forms resides in their DNA, which is physically a *part* of the forms specified, that of the HP forms is more a property of the forms than a part of them, and manifests as the mental conditioning of the individuals who carry (or carry out) the forms in question. But that human affairs form a category apart from anything in biology is a universal belief – insect societies, which perhaps come nearest, fail in autonomy, since this is determined by the genetic inheritance of the species and not by any feature of the social organization itself. Moreover the man–animal difference is felt as being of the same kind and degree as the living–inorganic difference. But the HP unlike the OP is not fully within the competence of science, for this is itself one of its forms and so cannot 'stand out of its own light' as it can in relation to the PP or the OP.

3.B. THE INCHOATIVE PLANE

However we choose to describe the phenomena in question, no one doubts the existence of the things which I have assigned to the PP, the OP, and the HP. I shall however postulate the existence of yet another plane, whose contents, I believe, have been universally overlooked. This new plane I call the Inchoative Plane (IP), and it is supposed to underlie the PP in the same manner that the PP underlies the OP. One must therefore expect that it will be essentially simpler than the PP, and I give a firm mathematical meaning to this idea by postulating that the theory of indistinguishables, when it is fully developed, will give a sufficient description of the IP to account for its relation to the PP; and, we may hope, there will be consequences of it observable in the PP itself.

This hypothesis requires, by the analogy of the cases mentioned above, that physical entities are autonomous fluent forms manifested in the substance of the IP. If we accept, as part of the hypothesis, that they are at least *forms* on the IP, they are either statant or fluent; if statant, each such form is composed always of the same collection of indistinguishables, to assert which of any two observations of the form entails that we recognise the indistinguishables as the same ones each time, which is contrary to the nature of indistinguishables as discussed in Chapter I. Therefore physical entities must be fluents, though with the gloss that in this plane there are no other forms than fluents.

But, if they are fluents, in this rather metaphysical fashion, how do they retain their self-identity, as many of them seem to do? This is the property which physicists call 'stability' (though commonly qualified by a parameter of duration, such as 10^{-23} sec.). It is supposed to be a physical property, that is, it exists *within the PP*, and it is therefore by the definition I have used above an instance of autonomy of the forms concerned. These are then autonomous fluents on the IP. It follows that, *if* there is an IP underlying the PP, then the PP has the same kind of relation to the IP that the OP has to the PP or the HP to the OP. And if there hangs an aura of triviality over the matter, isn't that what one would expect in such a case, where it is likely that we are nearing the frontier of scientific competence?

In science we look in general for explanations in terms of causal relations. There is an obvious pattern here in terms of the planes I have been describing; namely that each plane constitutes a set of constraints or limiting conditions on the next. In the prototypical case, the PP constrains the OP; that living organisms are physical bodies is so obvious, indeed, that it has historically been difficult to grasp that they are not *merely* bodies of this kind. It is also accepted that human society is constrained by the animal nature of men, so that how we get enough to eat is a major factor in all our affairs, and so on. Though there has never been any analogous difficulty in seeing physical entities as things in themselves, not merely forms in some lower plane, the analogy indicates that whatever can be established about the IP must constrain the PP; that is, that *some* physical 'laws' ought be traceable to the IP, if this can once be described well enough to see how. On the other hand, just as there are objects in the PP which owe their existence to the OP (e.g. stromatolites, anthills) and which therefore must remain unexplained within the PP itself, so there ought to be IP features which are not explainable in its own terms. But we cannot predict that organisms *will* exist from physical principles, but at best see that they *might* exist; the OP is not reducible to the PP in a strictly reductionist causal manner. Likewise the PP is unlikely to be reducible to the IP, for it will be full of things understandable only in its own terms.

Such considerations turn out to have a determining influence on the way the overall theory which I shall present is put to the test in its capacity as an 'explanation' for certain physical facts. I shall therefore devote separate sections to the questions concerning observation of the IP and the recognition of its effects on the PP.

3.C. OBSERVABILITY OF THE INCHOATIVE

Observation is a term describing a physical process. By some succession of causal encounters, terminating in the brain of an 'observer', information reaches the observer concerning (as it seems to him) something existing in the physical world. There is no analogue of this process at the biological level close enough to be of any help in generalizing the observational process from the PP to the OP. Can it be transferred, on the other hand, to the IP?

The problem may be put in this way: given two entities X, Y, to devise a sequence of operations which, if feasible, would result in discriminating all three cases allowed in a triparitous theory. To keep the matter within the terms of the informal definition (1.1), I shall deal only with classes and operations on classes. I have given the proof that we can in fact proceed from (1.1) to a more rigorous definition in the last chapter. The things we are allowed to do by (1.1) are to form classes and to determine their cardinality; there is however one counter-intuitive detail, namely that *two* identicals, whether classes or individuals, cannot be distinguished from *one*, because if they were not one they would not be identical. Subject to this restriction, we may do anything which can be clearly defined within the terms stated.

If we form the class [X, Y], we can tell at once whether $=$XY, for if so we shall have a singular class, and if not, not. But we are no nearer distinguishing $\hat{-}$XY and $\dot{-}$XY. It is clear from (1.1) that we need to form more than one class to make any progress. So let us take two arbitrary entities A, B and form with them a class O = [A, B]. I shall assume that O is well-behaved in the sense that there will be no difficulty in operating with it and recognising it again afterwards; this can *not* be assumed for X or Y. Thus, if we next form the class [X, A, B], and then subsequently try to form [Y, A, B], there's no knowing what we shall get; for if $\hat{-}$XY the one we pick second time may or may not be the same one as we got first time. So the proper procedure is to perform the operations *simultaneously*, which I shall assume to be possible.

Thus, given A, B, X, Y we form in one operation a class R of three classes $R = [O, P, Q] = [[A, B], [X, A, B], [Y, A, B]]$. We can assume $\underset{-}{!}AB$ because these are arbitrarily chosen, and to avoid duplication of cases I shall assume that if either of these is identical to X it will A rather than B. There are then seven possible cases:

(1) If $=XA$, $=YB$, $\underset{-}{!}XY$ then $O = P = Q$ and ${}^{c}O = 2$
(2) If $=XA$, $\ddagger YB$, $=XY$ then $O = P = Q$ and ${}^{c}O = 2$
(3) If $=XA$, $\ddagger YB$, $\underset{-}{\hat{}}XY$ then $O = P < Q$ and ${}^{c}O = 2$, ${}^{c}Q = 3$
(4) If $=XA$, $\ddagger YB$, $\underset{-}{!}XY$ then $O = P < Q$ and ${}^{c}O = 2$, ${}^{c}Q = 3$
(5) If $\ddagger XA$, $\ddagger YB$, $=XY$ then $O < P = Q$ and ${}^{c}O = 2$, ${}^{c}Q = 3$
(6) If $\ddagger XA$, $\ddagger YB$, $\underset{-}{\hat{}}XY$ then $O < P \underset{-}{\hat{}} Q$ and ${}^{c}O = 2$, ${}^{c}Q = 3$
(7) If $\ddagger XA$, $\ddagger YB$, $\underset{-}{!}XY$ then $O < P \underset{-}{!} Q$ and ${}^{c}O = 2$, ${}^{c}Q = 3$

In the first two cases our operations yield only one class, and thus no information. If in these cases however we replace R by R′ differing in that A is replaced by a new item distinct from it, say C, then either we shall in repeating the operations find a different case or we shall do so on similarly replacing C by a third choice D. The next three cases (3, 4, 5) yield two classes, but cannot be discriminated because P might be either and cannot be separately recognised on account of such identity. Cases (6, 7) yield three classes, including a non-identity between P and Q, which may be either twins or distinct, according to the initial parity distinction between X and Y. The threat of infinite recursion can be averted at this point by forming a new class $S = [O', P', Q'] = [[O, P], [P, O, P], [Q, O, P]]$ and repeating the same operations, which will now necessarily be confined to cases (3, 4), will yield two classes, but again fail to discriminate between $\underset{-}{\hat{}}XY$ and $\underset{-}{!}XY$. Thus, by exhaustion, these two cases cannot be separated.

To the reader who objects to this argument on the ground that it rests on a model of 'observation' barely related to what real observers really do, and therefore that its failure to discriminate indistinguishability from distinction should rather be adduced against the use of the definitions (1.1), I answer that, firstly, a coherent theory *can* be derived from (1.1) if *indistinguishables* are not empirically detectable as such; and secondly that this must be established by considering observational techniques apposite to indistinguishables as thus defined. Moreover, when the same procedure is applied to strictly biparitous objects, all cases reduce, or can be converted to, (5) and (7), which do clearly discriminate identity from distinction. A further possible objection is that by direct observation we could tell if $\underset{-}{!}XY$, and if this were not the case, then the indirect procedure described would establish whether $=XY$ or $\stackrel{\wedge}{-}XY$. But if $\stackrel{|}{-}XY$, we could not carry out the initial step of the procedure, since the identity of X, Y would be lost, and we would not know whether P, Q contained different or identical members of the pair; in fact, the *labels* [X] and [Y] are only *defined* by the initial construction of the class R, and cannot be compared with some previous use of the same names *unless* it is found that $\stackrel{!}{-}XY$; nor can one succeed by detecting directly a distinction between P, Q at the end, for being already members of the one class R, their non-identity holds whether X, Y are twins or distinct. We are therefore obliged to conclude that

> Empirical observations provide no legitimate occasion for the use of triparitous terminology. (3.0) Th.

We can now classify 'objects' according to their observability; but before doing so I ought to point out that the notion of an 'object' itself is far from intuitively obvious. I contrast objects with events on the one hand and attributes on the other. Events can be at least approximately localized in space and time, and can be said, albeit in everyday contexts pedantically, to be the only true observables; objects are recognised by their participation in or effect on events; and attributes can be defined or perceived only in relation to some object possessing them. I refer throughout to objective events only, though of course subjective events (objects, attributes) also exist; on the other hand I consider here both 'macro' and 'micro' events. Here are some examples of events, objects involved, and attributes of the latter: An *impact* may involve a *cricket-ball* which has *inertia*; a *snapshot* involves *light* which has *colour*; the *deflection* of a passing body may be due to a *black hole* which is *invisible*; the *decay* of a hyperon requires *space* which may have *intrinsic curvature*; a flash of *lightning* involves *electrons* which have *charge*; the *crystallization* of caesium involves *atoms* which can be identified by their *spectrum*; a *falling apple* is affected by the *gravitational field* which obeys locally the *inverse-square law*.

Now every object, as thus defined, must be *either* (a) a member of a class of indistinguishables, *or* (b) an aggregate of indistinguishables, *or* (c) composed (directly or indirectly) of such aggregates, *or* (d) inherently biparitous, in the sense that there is no object indistinguishable from it or from any part of it. As we have seen, class (a) objects are not observable individually. Those of class (b) can be involved in momentary simultaneous observations, but since their constituents are unobservable it is meaningless to postulate that two successive observations refer to the *same* such object – at best we could be observing, on the second occasion, a similar one. Coming to class (c), however, we have their internal structure to consider, and if this is encountered in successive observations, and involves similar constituents in corresponding roles, we can attribute individual identity to the objects observed.

Class (d), though we commonly think of all objects as belonging here, is more questionable. Objects whose description is wholly mathematical, such as the gravitational field, might be included – but these might perhaps be better classed as attributes, in this case of space. I shall argue in due course that as far as localizable entities are concerned, if not absolutely, this class may well be empty. This cannot of course be proved; but it is an interesting speculation.

We conclude that direct observation of the IP, defined as hypostatical to a triparitous theory, is out of the question. Observation, in fact, is definable only in the PP and is only relevant to the PP. The only way to get evidence for the existence of an IP at all is therefore to consider what logical conditions would enable us to infer that some 'law' of the PP was in fact a constraint on the PP originating in the IP. In this way, we keep observation within the PP, but (if the conditions can be met) can make a case for involvement of the IP in an explanatory capacity.

3.D. METHODOLOGY

Suppose that some observed feature of the physical world is suspected of being explicable as a constraint emanating from the IP: what steps can we take to assess this hypothesis? Of course, as with all hypotheses, it cannot be proved; it may be refuted, but if not it remains no more than plausible, though with enough such cases, and no straight contradictions, the theory as a whole may be rated higher.

To gain plausibility for such an explanation we must first find some feature (or set of features) of the IP which the observed facts might be said to reflect. This can come only from an inspection of the theory, since nothing in the IP can be directly observed; and it is a matter of creative searching rather than a rule-bound procedure, though the repertory of 'features' provided by the theory is finite, and small enough to allow of final failure. Having found some posssible basis for the supposed constraint, we have next to construct a model or description of it in terms logically capable of describing physical observables, while at the same time deriving from the (in general) triparitous description of the inchoative object and imaging it appropriately in biparitous terms.

That is to say, starting from a triparitous description, and not introducing any material not present therein (this assumes that the effects we are seeking to explain are *solely* due to IP constraints) we must construct a *biparitous homomorph* of the given inchoative object. Failure of homomorphism means simply describing the wrong thing; and failure of biparity means not meeting the demands of empirical validation. The set of constructs arising from the theory of indistinguishables, and therefore putatively existing in the IP, which can be successfully handled in this way is small, spiky, and suggestive. But it will not emerge before a great deal of theory has been gone through.

Before going on to reduce the above ideas to a formal definition of properly interpretable, or as I shall call them *rational* inchoative systems, it will be helpful to consider next what kind of physical observations may be candidates for this kind of explanation.

3.E. TYPES OF INDISTINGUISHABLES

One of the most general mathematical ideas used in the description of physical entities is that of a 'vector'. This is usually defined as an ordered set of elements of a given 'scalar' algebra, subject to certain additive and multiplicative operations. Strictly speaking a vector is not *meaningfully* an ordered set, in that it is only a labelling of the components that is involved; among themselves the components are usually in complete symmetry. Suppose now that we have a set of objects, of which all we at first know is that each can be associated with say a real 3-vector. In this situation we shall want to know what these vectors mean in physical terms; we shall be answered, if we can identify (or label) the vector components as referring to some set of three coordinates, and then, for each object in the set, make measurements to determine the values of the coordinates proper to it. Before making these measurements, which we predicate of an object, we already have an interpretation of the vector in terms of the system of coordinates; before making this interpretation, we have only a set of three indistinguishables.

Most readers will have been thinking here of the interpretation as in 3-space; but this is one of many possibilities. The values assigned to the vector components need not be numerical, or may be multiply-numerical; we could be assigning 'flavours' to three quarks, for example.

If we start with a larger class of indistinguishables, seven perhaps, we could interpret them as quantum numbers, collectively defining a flavour, of which can be predicated only sets of discrete values. As soon as any *physical correlation* is arrived at, we part company from indistinguishables and deal instead with a set of distinct descriptors (to use the most neutral term); as soon as we *predicate* particular values for them, we place them in some kind of metric space.

When we come to look at larger classes of indistinguishables, we shall find usually that the set of values which may be predicated of each gets smaller, ending with only two, which are usually picturable as describing existence as against non-existence. In such a case, we have to do with a set of *objects* of some kind whose cardinality has a fixed upper bound, given by the cardinal of the indistinguishables. There is no clear-cut distinction between an interpretation by vector components and by objects which may be present or absent from a given set. Quarks for instance can be treated for some purposes as vector components whose values are flavours, while for other purposes they may be thought of as objects (as in Quantum Chromodynamics). It is convenient to make a verbal distinction between correlations with objects and with components; but it denotes no sharp dichotomy.

As between these two, components are clearly the more fundamental; it is easier to see the presence or absence of an object as a special case of evaluating a component, than it is to see a component capable of a notional continuum of values as a funny sort of object. The ultimate building blocks of the universe (to use a misleading expression) are more like numbers than like things (which we are misled into thinking of them as). But, if my hypothesis is given any more than an instrumentalist accord, they are really neither of these: they are the indistinguishable forms of the inchoative plane. (But don't ask 'Forms in what substance?' until there is some consensus that the forms in question really exist – for they may not.)

Whatever kind of correlation we may consider, the primary datum which carries over from the IP to the PP is always a number, or small set of numbers. It is therefore numbers which will be the primary evidence for the correctness of the present theory. This is in one way to be expected: the reason why the various dimensionless ratios appearing in fundamental physics have the particular values they are observed to have has always been a puzzle. There is hope then that these ratios, or others derivable from them, might find an explanation in terms of 'constraints' imposed on the PP by the IP. But in practice we find that most of the numbers predicted are small integers, whose evidential value, each on its own, is negligible. The situation is saved from triviality by a few large or non-integral predictions – and also by considering how the total explicans furnished by the theory fits the total explicandum of the physical world.

3.F. THE PRINCIPLE OF COHERENCE

Logically, my initial postulate concerning the inchoative plane entails that every ostensible constraint computable from it should be observable in some numerical feature of the physical plane. It is allowable to suppose that there are a few instances still awaiting discovery, but they must be very few to carry any conviction in favour of the hypothesis. If any one is clearly in error, the whole theory must collapse. Because of this, the more tests the theory survives, the stronger every item already confirmed becomes, even those having little weight by themselves. It is no great feat to 'predict' that space should have 3 dimensions, but rather smart to get the fine-structure constant right to 5 decimal places. Nevertheless, if the former were incorrect it would be just as disastrous for the theory as if the latter were (it probably *is* a bit out: but it's a lot easier to make excuses for an error of 0.001% than for one of 30% !). Thus the fact that our conclusions come as a complete set of two dozen or so, which must show up simultaneously in this way, all standing or falling together, makes the theory more than trivially testable.

Indeed, this principle of coherence has more to it than mere numbers of results. The set of 'rational Sorts' we come up with has a distinctive pattern or structure, which can be correlated with an analogous pattern among the items selected as interpretations; and of course can be advanced in support of the particular correlation proposed, for it is easy to show that that 'wrong' ones would destroy the pattern.

We find ourselves then with a rather unusual situation. For on account of this coherence of the entire exercise, I have very little room for manoeuvre if it should happen that anything comes out wrong. One can always make mistakes in calculation, of course; and later correct them. But once this is done, and confirmed by others, there is no opportunity (or almost none) to save the day by minor adjustments to the theory. This makes it particularly difficult to assess the results we do find. For, as with any other such enterprise, success is less than complete: there are a few rough edges and doubtful correspondences. But, I hope to show, so few in comparison with the successes that it is hard to dismiss the latter as mere coincidence.

Despite the importance of coherence in testing the theory, it is not a matter which can readily be quantified, nor can it be stated as a rule constraining any particular step in the development. By contrast, the conditions to be met by individual IP systems in order to qualify for empirical status are very readily reduced to the form of rules, which will be repeatedly referred back to hereafter.

3.G. VALID REPRESENTATIONS

I have explained, in Section 3.D, that to have any empirically observable correspondent, an inchoative system must be biparitously describable in terms derived entirely from its own material, so that a biparitous system can be *derived from* the triparitous datum which is also *homomorphic* to it. (It clearly does not matter if a given triparitous entity maps onto two distinct elements of the proposed representation, since redundant elements can always be discarded; but if two entities map onto one element, the representation is obviously inadequate.)

These conditions can be stated more formally as follows:

If D is an inchoative system having an empirical counterpart, the latter must be adequately modelled by a valid representation H of D, satisfying the conditions that:

> H is biparitous, and consists of formulae of D. (1)
> There is an injective mapping from D to H by elements. (2)
> There is a bijective mapping between all functions definable in D and some set of functions definable in H (any others being disregarded as part of the representation). (3)
> The element representing an element of D having twins must have no pertinent singularity as against those representing such twins. (4)
> Every relation constitutive of H as thus understood must also be a relation in D with the same truth-value. (3.1.5)

The points (1) and (2) come directly from the specification set out in Section 3.4; and (3) merely applies the requirement of homomorphism specifically to functions. (4) is a necessary, but perhaps not sufficient, statement concerning homomorphism in regard to the elements of D and H. It entails that over any subset H′ representing a part D′ of D all of whose elements are indistinct, any relation constitutive of H must be an equivalence. Finally (5) indicates the immediate use to be made of the fact that any relation in H must be also a relation in D by virtue of (1); as stated, it is necessary so as to avoid (for instance) taking $= x\,y$ in H to 'represent' $\neq x\,y$ in D. (5) also forbids taking the inequality between different elements of H representing one element of D, as allowed in (2), as a relation 'constitutive' of H; this can obviously be done by defining H to exclude all but one of any set of co-images. If we assume that this has been done, and incorporate it somewhere in the rules, the condition (3.1.5) can be stated more briefly.

For the purpose of checking out a suggested representation for conformity with the conditions, it is convenient to state these conditions at somewhat greater length and with reference numbers attached. What we get is this:

R: If D is a triparitous domain
that is: a class of elements over which certain functors can be defined and within which at least some pairs of elements are twins
it is said to be *rational* if and only if there is a valid representation of D in a domain $\hat{\mathrm{H}}$
that is: at least one subset H of $\hat{\mathrm{H}}$ such that

R1 : in the subset H, every
- **R**1.1 : element is expressible as a formula in **T** using only symbols definable in D; and
- **R**1.2 : pair of non-identical elements
  - **R**1.2.1 : are distinct in D, and every
  - **R**1.2.2 : are images of non-identical elements of D; and
- **R**1.3 : relator
  - **R**1.3.1 : governs a relation which, if true, can be asserted in both H and D
  - **R**1.3.2 : and is an equivalence over any set of elements of H which are images of twin elements of D; whereas

R2 : in the domain D, every
- **R**2.1 : element is represented by one element of $\hat{\mathrm{H}}$ and only one element of H; and every
- **R**2.2 : parity relation between elements is determinable from formal properties of their images in H; and every
- **R**2.3 : functor is represented by a functor defined over H in terms of functors (not necessarily the same) defined over D (3.2)

3.H. THE CONSTRUCTION OF REPRESENTATIONS

It will be clear from all that has been said in this chapter that the application of the theory, described in Part II, will depend entirely on the complete and correct identification of the 'rational' inchoative systems as defined in (3.2). Accordingly, my exposition of the theory will not follow the usual pattern of a pure mathematical treatise, but will be in effect a one-way road to this specific problem. I shall now briefly outline the strategy which I shall follow in this; this strategy will be worked out explicitly in Chapter VII.

The principal constraint is that imposed by R1.1. Our representations are to be constructed out of the elements and functions definable within the systems to be represented. I shall prove, perhaps at first surprisingly, that formulae defined in terms of elements which are all twins *can* denote distinct entities. Therefore if we collect all the formulae definable with a given functor, we shall usually find that some are twins and some distinct (as well as some identical). Owing to the limited variety of functions definable, it is not impracticable to survey the entire field of possible formulae, and to find in particular the largest subclasses which are ordinable (that is, have every pair of elements bipar) in each class.

The next step, having now satisfied R1.1 & R1.2.1, is to eliminate those ordinable subclasses which do not satisfy R1.2.2. The major source of representations, we shall find, is happily such as to make this step especially easy. When we come to R1.3, things get a bit more critical. R1.3.1 it turns out cannot fail to be satisfied provided R2.3 is so. It might seem very laborious to check all the functors of a system D, but they turn out to be very few, and there is, among the critical cases, only one functor which turns out not be representable at all. R1.3.2 on the other hand, trivial for very small domains, gets rapidly out of hand, but not before we are able to prove that it will not work for any domain larger than a certain cardinality. Though stated in terms of equivalence-relators, it is essentially a symmetry condition and is satisfied, for instance, for any set of elements between which no relations at all can be asserted in H.

The second part of the test, R2, is largely a question of being able to make appropriate selections from an initial $\hat{H}$ to give a valid definitive representation H. Only R2.3 ever gives real trouble. Nevertheless it is essential in each case to check that R2 is in fact satisfied throughout.

3.J. THE IRREDUCIBILITY OF THE PHYSICAL PLANE

The relation between the PP and the OP, and again that between the OP and HP, is unsymmetrical. Living organisms are entirely subject to the constraints of physical laws, since their bodies consist of physical matter; but for all that, as a result of biological processes, they have acquired many features and functions which cannot be fully explained by physical or chemical principles, even though they are of necessity compatible with them. This may be put more briefly, by saying that the OP is not *reducible* to the PP; those who deny this are called 'reductionists', and if their view were sound (as in many other instances it is) the two planes would reduce to one. Likewise, the HP is not reducible to the OP.

Does the same irreducibility apply to the PP? In Chapter XIV, where the results of the theory are discussed in detail, it will appear that the contribution of the IP, if it is at all completely described by the theory of indistinguishables, is manifold but patchy: there are many points in fundamental physical theory which are not touched. No doubt some of these will be brought into relation with the IP by future research, but it seems unlikely that all will be. The prospect is that the IP will not explain everything – (if it did, it would be an *a priori* theory of creation itself) – which of course is what the analogy of the other planes leads us to expect.

This analogy fails us however in another respect. For whereas the irreducibility of the OP is partly concealed by the fact that the PP is fully observable to us, and so constantly at hand as an explanation for various problems of the OP, the unobservability of the IP precludes (or, at least, has hitherto precluded) any comparable getout for problems arising in the PP. It is in response to this situation, I would like to suggest, that physicists have built up a considerable body of theory intended to incorporate the constraints, due (in my view) to the nature of the IP, into a fairly coherent physical theory. That is to say, a kind of inverted reductionism is observed, whereby as much as possible of the IP is 'adopted', as it were, into the PP.

If the insights on which the present work is based are to be trusted, this procedure is unlikely to be ultimately successful. One would hope that a complete and mathematically rigorous theory of physics would eventually emerge, based on axioms which could be justified, if such justification were called for, by appeal to the very much simpler theory of the IP, of which the present work is a first draft. If so, we need to have some means of making the desired connection.

3.K. THE COMBINATORIAL HIERARCHY

I have argued above that any triparitous class which has of itself any discoverable empirical effects must have a valid representation of the type defined in (3.2). Merely to find a biparitous homomorph of such a class tells us nothing about it, since it is always possible. For example, the elements of the class D could be represented by integers of the form $p2^n$ where p is prime, and n non-negative, numbers with the same value of p being taken as representing twins in D; any functions required could be defined by arbitrary operational tables on these integers.

However, this does not prove that a biparitous representation of such a triparitous class D *not* derived from the elements of D has no validity; for if D also has a 'valid' representation, then by the principles of biparitous mathematics any other system isomorphic therewith is equally available for use. Thus, once we have proved that the class of two indistinguishable members (the smallest non-trivial case) which I shall call $D2_0$ is a 'rational Sort', we are free to look for representations of it other than the $H2_0$ which I develop, in accordance with the principle of this chapter, in chapter VI. It is only necessary that such secondary representation should embody at some point a feature of total symmetry involving just two elements.

The conditions are met for example by a system of 2-vectors over the field J_2. The twoness of the vector components can be correlated with the two members of $D2_0$, as they are completely symmetrical (but *not* with the two elements of J_2, which are not symmetrically related). However this scalar algebra does provide *labels*, which can be predicated of the two vector 'slots', giving (1, 0) and (0, 1) as two initial vectors which *indirectly* represent the elements of $D2_0$; their sum (1, 1) could then represent the blur of the two elements, completing the representation of the closure $\overline{D2_0}$ by the linearly independent set of three vectors. I use here terminology which will be explained in Chapter V.

It is from the above starting point, though under different terminology, that Bastin *et al.* [4] construct their 'Combinatorial Hierarchy'. This is put forward by them as a physical theory, that is as pertaining to the PP exclusively, and this claim is clearly justified in that none of the presuppositions of the theory of indistinguishables which prevent its immediate interpretation in physical terms are required to expound such a system. The system of matrices and vectors used in the Hierarchy are analogous to the pair-trees used in my representation developed in Chapter VII;

but at no point do they have roots in a triparitous theory, and are axiomatized in purely physical terms.

Any property of such a system is in principle a candidate for a physical interpretation, and a number of suggested interpretations are given in the paper cited. It constitutes a logically independent theory, and any common ground among the interpretations arrived at between the Hierarchy and the present work lends mutual corroboration to both. However, the independence of the two theories makes possible the emergence of cases where a given phenomenon is linked by them to a point in their theory seemingly non-parallel to that which in mine is held to constrain the PP to exhibit the same phenomenon. Unless subsequently resolved, such an instance is one of *over-determination*; this is not logically precluded, but on any reasonable expectation should be rare. More than a few such cases would impugn the principle of interpretation used in one or the other theory.

Nevertheless, the use of such secondary representations – to use the term natural for them from the point of view present here – provides a legitimate stepping-stone by which to cross from exclusive concern with the IP, which can only propose constraints on the PP and not predict without physical considerations precisely what form they will take, to an equally exclusive concern with the PP, which must rely (I presume) on many axiomatic statements which, it is reasonable to hope, would find rational explanations in a theory of indistinguishables.

CHAPTER IV

SORT THEORY – AXIOMS AND DEFINITIONS

I have now explained the basic idea of 'indistinguishability', and expounded briefly the necessarily somewhat novel semantic theory required to design a proper notation in which we may discuss indistinguishables with mathematical rigour. I have also examined in some detail the logical problems arising from any attempt to link empirical data with a hopefully explanatory theory, in the case where the latter involves indistinguishables of any kind. Guided by the immediate goal, suggested by this analysis, and using the notation $\mathscr{T}$ with the rules of interpretation given in Chapter II, I start now on the systematic exposition of one theory of the triparitous type.

This will be the theory **T**, and will take the form of an analogue of Set theory in a triparitous context. Since as we shall see the term 'Set' would be thoroughly misleading, if extended into such a context, we need an alternative. If one is given a miscellaneous collection of objects, between which one is required to observe certain distinctions while ignoring others, by arranging them in appropriate classes, what one then does is called 'sorting' the objects. What I have called 'indistinguishables' answer to such objects, except that the distinctions to be ignored are precisely those that are unobserved and indeed radically unobservable. So I shall call **T** *Sort Theory*. The terms 'Sort' and 'Set' will be used similarly but never interchangeably; being both common words, they will be capitalized.

4.A. INDEFINABLES OF T

Sort theory uses five formally indefinable symbols. Three of these are the parity-relators $=, \hat{-}, \dot{-}$ whose usage is governed by the Axiom Schema (4.2). The fourth is $\in$ denoting class-membership, so that $\in Cx$ is true if and only if x is a member of the class C. It is subject to the Axiom (4.1) below. The fifth is **C**, appearing only in the compound expression $\in\mathbf{C}$ which predicates that its sole argument is a class; such relations play a part only in a few initial statements, and the anomalies liable to arise from the general notion of a class will not concern us.

The first Axiom of **T** sets this theory off sharply from Set theory by making the domain of $\in$ coextensive with **C** so that there is nothing which does not have members:

$$\in \mathbf{C}x, \in \mathbf{C}y, \in xy, \quad \in yx \Rightarrow = xy \qquad \text{(4.1) Axiom}$$

The protasis is written in analytic form, but can be reduced to canonical by conflating the first two terms using $\hat{\cup}$ and using the notation of 2.J. Note that identical classes are not necessarily mutual members.

There follows an axiom-schema on substitution of symbols, which formalizes the conclusions reached in Chapter II about the semantics of triparitous theories. The new axiom schema replaces the one which holds for (so far as I know) all biparitous mathematics, and is in consequence rarely made explicit, though it appears in Bourbaki [3], Chapter I; inevitably, the new rules are more complex than the old, which merely authorize the substitution of equals for one another, and forbid the same as between unequals. The rule for the present theory is:

> If $P \Rightarrow A$ is any statement in $\mathscr{V}$, with P, A being symbol-strings obeying the syntax of $\mathscr{T}$, two of whose symbols are denoted by x, y, then
>
> $P \Rightarrow A'$ has the same truth-value as $P \Rightarrow A$, where A' is formed
>
> if $P \Rightarrow = xy$, by replacing any occurrence of x in A by y, *or* of y by x (4.2.1)
>
> *but* if $P \Rightarrow \hat{-}xy$, by replacing every occurrence of x by y, *and* of y by x, throughout any nexus in A in which any one such replacement is made (4.2.2)
>
> *and* if $P \Rightarrow \nmid xy$, by replacing every occurrence of x by y, and of y by x, throughout P and every relation inferred from P (4.2.3)
>
> whereas if none of these implications holds, no replacement involving x or y is hereby legitimated (4.2) Ax. Sch.

The novel element here is of course (4.2.2); the term 'nexus' is defined in (2.27). The main part of the difficulty of Sort Theory arises from the unfamiliarity of this substitution-rule, and the laboriousness of checking each step in an argument for conformity with it. I shall now give a method and some examples of its use in this process.

4.B. METHOD OF VERIFYING CONCURRENCE

When faced with a complex expression, in which one has to verify the statements made about concurrence of symbols, and comments thereon, one method is to write the expression in the form of a tree diagram. The use of preposed functors greatly assists in this. The procedure is to write down the governor of the expression (its first symbol) on the left, and draw lines from it as many as it has arguments. Then take each line in turn, writing in the argument, beginning with its governor if it has more than one symbol; when any argument is thus finished, do the same for any loose valency lines. Try it on the last protasis term of (4.12) (p. 65):

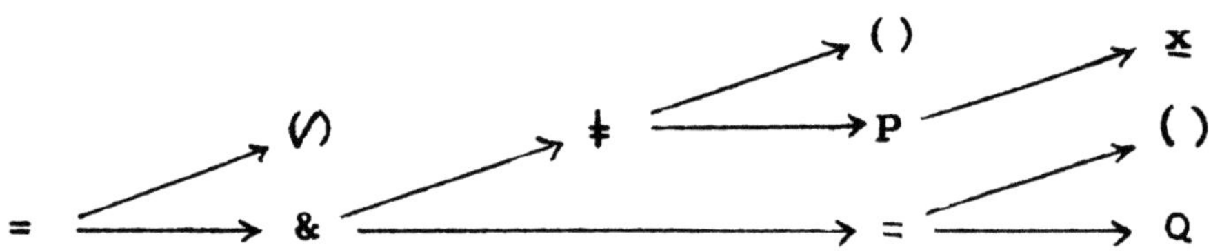

(Ex. 1)

The general governor here is =, which of course has two arguments to which we draw lines to the right from it. The first argument is the single symbol (/) which thus ends this branch. The second argument is written (ǂ( )Px = ( )Q); this stands for the conjunction of the and ǂ-relations, but the symbol in $\mathscr{T}$ for conjunction is null. Null symbols are inconvenient in tree diagrams, so I shall write it as &, with two valency-lines. The first line begins with ǂ which has ( ) as one argument and Px as the other; this last contains the functor P and the argument x, and is so displayed. The second governor, =, has two one-symbol arguments, ( ) and Q. This concludes the buildup of the tree diagram.

To find whether two symbols, as they appear in such a diagram, are concurrent, start with the leftmost, take one step back to its immediate governor, and then see whether there is any route forward from here to the other symbol which does not encounter a classifying functor, (with no declassifying governor) on the way. If there is such a route, the two symbols are concurrent (check this against the definitions in Chapter II). It will help at first to underline classifying functors before you start.

Let's try this on the statement (4.12), assuming that we can write A ⇒ B as '& A = (/)B, we get the rather formidable tree:

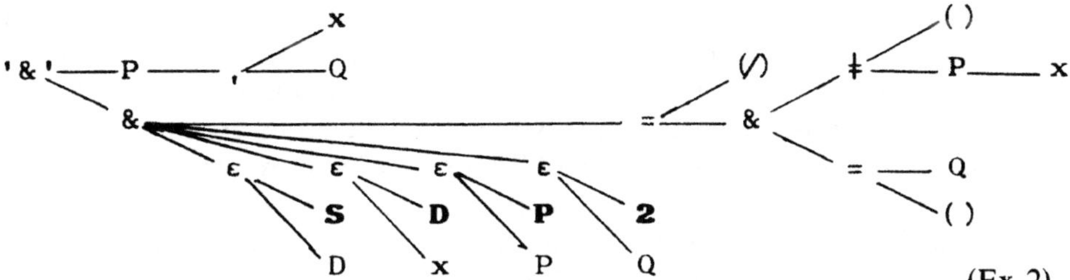

(Ex. 2)

The concurrences that need checking here are those involving the P's Q's and x's. As for P, we step back from the upper one to the governor '&' whence we can reach both the other two P's; none of our functors here are classifying. The x's by this test seem to fail but this can be overcome by writing ∤ (/)Px for ∈**PP**, so that x only occurs as Px, and is concurrent by rule C3 of (2.25). The test fails for the Q's also; in this case however we can rewrite the whole statement in canonical form as (4.12a), showing that in fact all the necessary concurrences can be secured. The reader may like to check (4.12a), remembering that *R and $^{+}$R count as single functors, where R is any relator.

Naturally, since statements involving non-concurrence entail the risk of ambiguity at least, we normally take care to avoid using them. It may happen however that the fact of such non-concurrence is itself of theorematic status, and in such cases the errant formula may be used in good faith, much as false statements are deliberately made in the course of proofs involving the *reductio ad absurdum* principle. An example of this is (7.4), which gives us the tree diagram

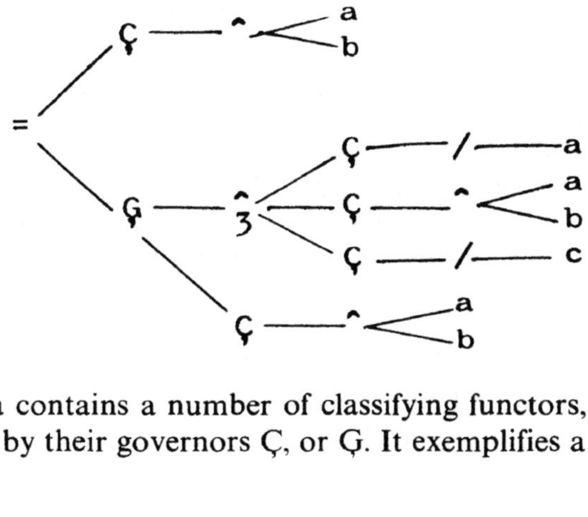

(Ex.3)

This formula contains a number of classifying functors, but they are all 'declassified' by their governors Ç, or Ģ. It exemplifies a simple but non-trivial form of a nesting of 'cedilla' functors, which are discussed in detail in Chapter VII. It will be observed that the symbol *a* occurs four times; the occurrence as argument under / cannot be reduced to concurrence with any of those governed by ˆ, while all of the latter are instances of a common formula Ç ˆ*ab*. In view of Theorem 2.24 moreover, the declassifying functor Ç can clear only one of following ˆ's so that the two lowest *a*'s are also non-concurrent, though the top *a* concurs with the bottom *a*. We conclude that the Equation (7.4) is in general irreducibly ambiguous.

If it were not for the classifying property of the functors /, ˆ however, Example 3 might be rescued in a manner illustrated by the last protasis term in (5.15.1) which is $= {}^{c}Y^{c}/,'' xx(\in ;; XY\, ''xx)$ and contains two occurrences of the formula $''xx$. Though $''$ is classifying, there is no declassifying functor involved and Theorem 2.24 is irrelevant; that all the four *x*'s are concurrent is shown by rewriting (5.25.1) as

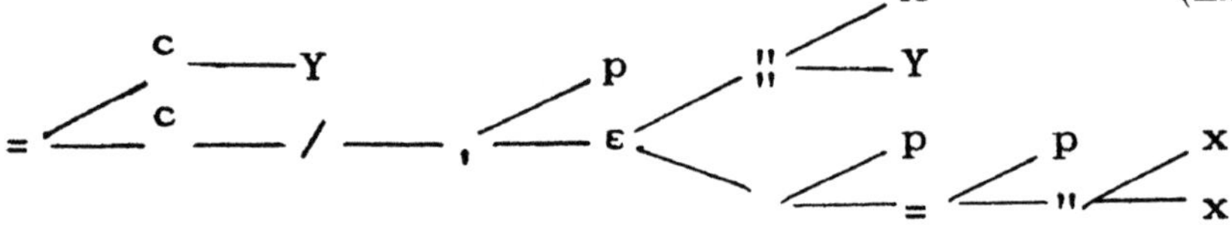

(Ex.4)

Here the two *p*'s are clearly concurrent, and only one need be equated to $''xx$, where the two *x*'s are also concurrent (note the use of , with singular second (predicate) argument). In such a case, the meaning of the formula is rather easier to grasp, when given in the form used in the text position cited, than in the above form; this is, as will be readily appreciated, an often valuable consideration – provided no mistakes are made!

4.C. DEFINITIONS IN THE INFERENTIAL SYSTEM

For completeness' sake it may be helpful to give here some definitions, and some very elementary theorems, which properly pertain to $\mathscr{V}$ rather than to the theory **T**. There is here only one indefinable, the implicator ⇒; consequently, the other implicators ⇔ and ⇒' need to be defined. The first is easy:

$$P,\quad R \Rightarrow Q,\quad Q \Rightarrow R \quad \Rightarrow \quad Q \Leftrightarrow R \qquad (4.3)\ \text{Df.} \Leftrightarrow$$

Note that what we here define is not absolute equivalence, but equivalence in the presence of P, though this might sometimes be null. It is usually advisable to state such general conditions of applicability as the P of (4.3), since in this relatively novel field they are apt to be less obvious than expected; in particular definitions of functors are often of limited validity, and so need to be expressed with $\Rightarrow$ rather than the equivalence adequate for identifiers; the essential matter of the definition is then placed in the last protasis term, which should be carefully checked for invertability with the apodosis (though it is occasionally allowable to waive even this relative equivalence). Thus, in (4.3), if the protasis is false, though nothing logically follows as to the truth of the apodosis, if it were in fact still true the protasis would have been wrongly stated; so the implicator is tantamount to $\Leftrightarrow$ though this symbol cannot be used since it has not yet been defined.

The other implicator $\Rightarrow'$ is used only in the definition of the negative superfunctor. A relation in **T** is an expression which may be either true or false (or undecidable), and thus carries in itself no implications (unless or until it is proved true); it is therefore not admissible for a relation by itself to appear as a protasis of a statement, or therefore on either side of an equivalence. I therefore define

$$=(/)\mathrm{R} \Rightarrow '\mathrm{R} \qquad (4.4.1)\ \mathrm{Df.} \Rightarrow '$$

$$=(\ )\mathrm{R} \Rightarrow ''\mathrm{R} \qquad (4.4.2)\ \mathrm{Df.}\ '$$

with the corollary that

$$=(\ ) \Rightarrow ''\mathrm{R} \qquad (4.4.3)\ \mathrm{Th.}$$

which incidentally expresses the fact that we are using classical logic as already stated in Chapter II. Another statement of this series is:

$$\nmid(\ )\mathrm{R}, \quad \nmid(/)\mathrm{R} \Rightarrow '\in \mathbf{2}\,\mathrm{R} \qquad (4.4.4)\ \mathrm{Df.}\ \mathbf{2}$$

All these are stated completely generally, i.e. with the 'P' of (4.3) null. (4.4.4) is not in canonical form, which may be written

$$*\nmid[^{+}\nmid[' \in \mathbf{2}\,\mathrm{R}]\,(/)\,\mathrm{R}]\,(\ )\mathrm{R} \qquad (4.4.4a)$$

which shows that the R's are concurrent; it also shows, by using the sufficient-condition superfunctor $^{+}$, that, as with (4.3), the definition is virtually an equivalence.

I next define a class of *predicates* as follows:

$$\in \mathbf{C}\,\mathrm{D}, \quad \in \mathrm{D}x, \quad \in \mathbf{2}\,\mathrm{P}x \Rightarrow \in \mathbf{P}\,\mathrm{D}\,\mathrm{P} \Rightarrow \in \mathbf{P}\,\mathrm{P} \qquad (4.5)\ \mathrm{Df.}\ \mathbf{P}$$

Here $\in \mathbf{P}\,\mathrm{D}\,\mathrm{P}$ means 'P belongs to the class of predicates decidable over D', and up to this point the definition is invertible, since if $'\in \mathbf{2}\,\mathrm{P}x$ then P is not decidable for some members of D. The second implication is a notational rider and is not invertible. The symbol **P** is chiefly useful in its second sense, and will be treated as an identifier and not (unless specially mentioned) as a functor. A useful corollary of (4.5) is

$$\in \mathbf{P}\mathrm{P}, \quad {}'\!\in \mathbf{2}\,\mathrm{P}x \Rightarrow {}'\!\in \mathbf{2}\,{}'\mathrm{P}x \qquad (4.6)\ \text{Th.}$$

I conclude this section on the inferential system with a list of elementary statements in $\mathscr{V}$ whose meanings form a subtly graduated series, most of shich will be used on occasion in the sequel.

ǂ( ) R ⇒ (/)	R is a tautology	(4.7.1)
(/) ⇒ R	R is a relation	(4.7.2)
( ) ⇒ ∈**2**R	R is a decidable relation	(4.7.3)
( ) ⇒ R	R is true	(4.7.4)
( ) ⇒ =( )R	R is true in the given context	(4.7.5)
( ) ⇒ ǂ(/)R	R is unrefuted (true or undecidable)	(4.7.6)
( ) ⇒ ǂ( )R	R is unproven (false or undecidable)	(4.7.7)
( ) ⇒ =(/)R	R is false in the given context	(4.7.8)
( ) ⇒ 'R	R is false	(4.7.9)
( ) ⇒ '∈**2**R	R is undecidable	(4.7.10)
=( ) R ⇒ (/)	R is self-contradictory	(4.7.11)

Of these, the first and last are of rare occurrence.

It is worth noting that of the logical functions possible for relations, only three (conjunction, disjunction, and equivalence) are unaffected if their arguments are or involve twins. The negations of the first two, and the inference function, are available for relations involving twins provided these are adequately controlled for concurrence. Counter-equivalence (the function that is true if the arguments are of opposite truth-values) and non-inference are not necessarily distinguishable.

4.D. DEFINITIONS IN T – BASICS

We now turn from the inferential system to the object theory itself and begin by stating certain basic definitions. In **T**, as I have already remarked, we do not have any analogue of the 'Axiom of Choice', and for this reason we are unable to develop any extensive theory of infinite Sorts, though as we shall see some do exist. This fact opens to us the possibility of replacing a whole string of independent existence axioms, such as we find in Set theory, by a single axiom-schema on definitions. For instance, there is an axiom of Set theory asserting the existence of a Set having two members; this cannot be proved from other axioms, nor be made one instance of some more general statement, lest the way be opened for contradictions to arise. In **T**, however, nothing untoward is derived from the following:

> The definiend of a valid definition in **T** denotes an entity which exists in **T** (4.8) Ax. Sch.

where a *valid definition* in **T** is any statement in $\mathscr{V}$ using well-formed relations of **T**, whose protasis contains no contradiction or any previously undefined symbol, and whose apodosis contains just one such symbol the *definiend*. For example, if we were to omit the third term of the protasis in (4.25), we could define a z such that $= \mathrm{P}z\ {}'\!\in zz$, which would make the protasis self-contradictory and thereby invalidate the statement as a definition.

We start with necessary but trivial definitions of the three parity-relations not listed as 'indefinables':

$$\in \mathbf{C}x, \quad \in \mathbf{C}y, \quad {}'\!= x\,y \Leftrightarrow \nmid x\,y \qquad (4.9.1)\ \text{Df.}\ \nmid$$

$$ {}'\hat{-}\, x\,y \Leftrightarrow \hat{+}\, x\,y \qquad (4.9.2)\ \text{Df.}\ \hat{+}$$

$$ {}'\dot{-}\, x\,y \Leftrightarrow \downarrow x\,y \qquad (4.9.3)\ \text{Df.}\ \downarrow$$

Let us now consider which of the above relators are *equivalences*. This class is defined as follows. First, in analytic form:

$$=(\ )\mathrm{R}xx, \ =(/)(\mathrm{R}xy, {}'\mathrm{R}yx), \ =(/)(\mathrm{R}xy, \mathrm{R}yz, {}'\mathrm{R}zx) \Leftrightarrow \in \mathbf{Eq}\,\mathrm{R} \qquad (4.10)$$

where the requirements of reflexivity, symmetry, and transitivity are fairly explicit; however, non-concurrence among indistinguishables is rife in (4.10), making it insufficient as a basis for inferences. The canonical form, which uses the functor , introduced first in (2.28), is

$$(\ ) \Rightarrow \in \mathbf{Eq}, \mathrm{R}(, \mathrm{R}(=(/)(\mathrm{R}x, y(', \mathrm{R}(=(/)(\mathrm{R}xy, \mathrm{R}y, z('\mathrm{R}zx)))yx))xx) \qquad (4.10a)\ \text{Df.}\ \mathbf{Eq.}$$

which in proper rigour is the form to be referred to in the following argument. Nevertheless, it's a lot easier to follow if you stick to (4.10) as given above; in effect, the existence of (4.10a) serves to license the use of the analytic form as if it were unambiguous.

I shall now prove

$$(\ )\Rightarrow \in \mathbf{Eq} = ,\ \in \mathbf{Eq}\,⸸ \qquad \text{(4.11.1–3) Th.}$$

Substituting = for the R of (4.10), its first term becomes $= xx$ which is true; the second truly declares that ($= xy$, ǂ yx) is false; and the third just as truly denies ($= xy$, $= yz$, ǂ zx). All three terms are thus true for = R = , whence from (4.10) ∈**Eq** = , Q.E.D.1. Substituting now ⸸ for R, we find for the first term ⸸xx which is true from (4.9.3); the second term denies that ⸸xy, $\overset{\prime}{-}yx$ and this denial is also obviously true; while the third term denies that ⸸xy, ⸸yz, $\overset{\prime}{-}zx$. But if x is not distinct from y, nor y from z, then x is not distinct from z, so that the conjunction is false, and the third term true; thus as before ∈**Eq**⸸, Q.E.D.2.

The remaining parity relators satisfy

$$(\ )\Rightarrow {}'\!\in \mathbf{Eq}\,ǂ,\quad {}'\!\in \mathbf{Eq}\,\hat{-},\quad {}'\!\in \mathbf{Eq}\,\hat{\div},\quad {}'\!\in \mathbf{Eq}\,\overset{\prime}{-} \qquad \text{(4.11.3–6) Th.}$$

First, ǂxx is false, whence (4.11.3) follows.

Next, $\hat{-}xx$ is also false, for since the x's are concurrent and indistinguishable as symbols, they denote identicals, and not twins, whence (4.11.4) follows.

Next, though $\hat{\div}\,xx$ is true, from (4.9.2) and the preceding argument, the conjunction $\hat{\div}\,xy$, $\hat{-}yx$ is obviously false, so the second term of (4.10) also passes, in the third term we can have $\hat{\div}\,xy$, $\hat{\div}\,yz$ but yet $\hat{-}zx$, for y could be distinct and therefore bipar from both x and z being twins, so that its denial is false, and (4.11.5) follows.

Lastly, $\overset{\prime}{-}xx$ is false, and (4.11.6) follows, Q.E.D.

4.E. THE CONDITIONAL QUANTIFICATION FUNCTOR

The functor was introduced in (2.28) in the course of discussing the problems of quantification in a triparitous notation. It was there argued that 'bound variables', as used in the conventional notation for quantification, should have no place in such a notation, because of their anomalous behaviour in respect of concurrence (though by reserving certain symbols for this use only, we could preserve most of the logic). This functor has not yet been formally defined in **T**, and this must be done before it can properly be used in subsequent arguments. I have in fact used it in (4.10a) by way of illustration of its use in proving the possibility of framing a statement so as to preserve concurrence of symbols, which more perspicuous presentation obscures; but elsewhere I have used only the necessary-condition notation for formal canonicalizations, arguably acceptable for (4.4.4) but hardly for (4.10).

It is however obvious that it is the latter notation alone which is logically acceptable in stating the canonical form of the definition of $_{,}$ in **T**. This definition is

$$\in \mathbf{S}\,\mathrm{D},\ \ \in \mathrm{D}\,x,\ \ \in \mathbf{P}\,\mathrm{P},\ \ \in \mathbf{2}\mathrm{Q},\ \ =(/)(\ddagger(\,)\mathrm{P}x, =(\,)\mathrm{Q}) \Rightarrow \mathrm{P}_{,}x\mathrm{Q} \qquad (4.12)\ \mathrm{Df.}$$

where, as in (4.3), it is understood that the apodosis or forms derived by substitution from it will not be used if any term in the protasis is false, making the implicator virtually an equivalence. The canonical form on (4.12) can be formulated thus:

$$(\,) \Rightarrow {}^{*}\!\in[{}^{*}\!\in[{}^{*}\!\in[{}^{+}\!\in[\equiv \mathrm{P}_{,}x\mathrm{Q} = (/)(\ddagger(\,)\mathrm{P}x, =(\,)\mathrm{Q})]\mathbf{2}\mathrm{Q}]\mathbf{P}\,\mathrm{P}]\,\mathrm{D}x]\,\mathbf{S}\mathrm{D} \qquad (4.12\mathrm{a})$$

The first two terms of (4.12) do not really constitute a 'necessary condition' for what follows; but since the arguments involved do not recur nothing untrue follows from the assertion that they do. The symbol $\equiv$ is the equivalence functor of $\mathscr{T}$ (not $\Leftrightarrow$ which belongs to $\mathscr{V}$); the availability of this functor in a triparitous context is stated in 4.C. Note that (4.12a) asserts a relation in $\mathscr{T}$ but not in **T**; its purpose is to demonstrate that P, Q, x are all concurrent, as does not appear in (4.12), so that the latter is a valid definition of $_{,}$. This need not be done in **T**; but obviously it is more convenient if we adhere to **T** for such statements, which, now we have defined $_{,}$, we can generally do.

In connection with $_{,}$ we have also to state the following axiom:

$$\in \mathbf{P}\,\mathrm{P},\ \ \in \mathbf{P}\mathrm{Q},\ \ \in \mathbf{C}\,x,\ \ \in \mathbf{C}\,y,\ \ \in \hat{}\,x\,y\,u,\ \ \in \mathbf{2} = (/)\,\mathrm{P}_{,}u(\mathrm{Q}u) \Rightarrow \hat{+}\,xy \qquad (4.13)\ \mathrm{Axiom}$$

$$\in \mathbf{2} = (/)_{,}\,\mathrm{P}(\in \mathbf{P}\,\mathrm{P})_{,}\,u(_{,}\,\mathrm{Q}(\in \mathbf{P}\,\mathrm{Q})_{,}\,u(\in \hat{}_{,}\,x(\in \mathbf{C}\,x)_{,}\,y(\in \mathbf{C}\,y)u)) \Rightarrow \hat{+}\,xy \qquad (4.13\mathrm{a})$$

the latter being the canonical form using $_{,}$. The axiom (4.13) forbids the use of negations of relations involving $_{,}$ with indistinguishable arguments which necessarily imply an existential quantification of the affected variable, contrary to the argument of 2.K. The functor $\hat{}$ is defined in (4.26).

The expression $\mathrm{P}_{,}x(\mathrm{Q}x)$ can be read 'P holds for any x such that Qx'. This is normally sufficient for a verbal understanding; but it can happen that the same symbol is qualified more than once by $_{,}$-expressions, as happens in (4.13a) with u, which is subjected to the condition Qu where this u is again qualified by $\in \hat{}\,xy\,u$. Such nested occurrences of $_{,}$ express simple conjunction of the imposed conditions, and the order of the nesting is usually immaterial (though it may involve other subformulae whose order does matter). We read the form P, *x*(Q, *x*(R*x*)) as 'P holds for any *x* such that Q*x* and R*x*'.

It is important to remember what is implied by P, *x*(Q*x*) in the case where Q*x* is false. = (/)(⫲( )P*x*, = ( )Q*x*) is true if either = ( )P*x* or ⫲( )Q*x*; so, if Q*x* is true, P*x* is true, but if Q*x* is false ⫲( )P*x* or = ( )P*x* may be true. Thus failure of Q*x* means that P*x* is wholly undecidable (unless from other information). Under no circumstances does P, *x*(Q*x*) imply that P*x*, even though this may in fact be the case.

Use of the functor , is by far the commonest occasion for the use of the auxiliaries ( ). It is well to remember that such symbols can be omitted without corrupting the syntax, as also can , . Thus, given the relevant domains, the symbol-string ≡ P, *x*Q*x* = (/) ⫲ P*x* = Q*x* (if presented as a complete formula) is unambiguous; first, ≡ must be followed by two relations, then , must have a symbol and a relation, here *x* and Q*x*, and it makes an identifier so that P, *x*Q*x* is a relation, and the rest of the string must be another relation, and so must that part after = (/); ⫲ must be followed by two relations, and P*x* is one and there follows = which might begin the second, except that = is meaningless if directly governed by ⫲ so we assume the second argument under ⫲, and as only one argument follows = that too has a null argument. Thus the form given as ≡ P, *x*(Q*x*) = (/)(⫲( )P*x*, = ( )Q*x*) can be reconstructed, and the null functor & recognised as well. It is however unwise to omit the auxiliaries.

The functor , can be used with ( ) or (/) as second argument. P, *x*( ) means that P is true for *any* value of *x* no matter what nonsense you think of. This form is rare, but it occurs in (4.18), which is recognisably a special case. P , *x*(/) is a well-formed expression, but means nothing whatever (but that it is a truth-value). The formulae , *x*( ) and , *x*(/) stand for 'any entity whatever' and 'nothing' (not restricted to be a truth-value); the latter might have a use sometimes.

4.F. DEFINITION AND CLASSIFICATION OF SORTS

We come now to define the central concept of **T**, a *Sort*:

( ) ⇒ ∈**S**, C(⫲( ) +̂ , *x*(∈C*x*), *y*(∈C*y*))   (4.14) Df. **S**

which asserts that any class is a Sort of which it is *not claimed* that *every* pair of members is bipar, i.e. either identical or distinct. Note that it may be claimed that *some* pairs are so, and it may even be the case that *all* are provided this has not been proved or asserted. The only classes which are *not* Sorts are those defined as such, or considered under an axiom (normally alas unstated) precluding the existence of twinship. We can further classify Sorts as *perfect* **R**, as *ordinable* **Q**, or as *mixed* **M**, according to the definitions:

$$\in \mathbf{S}, \mathrm{S}(= (\ ) \ddagger, x(\in \mathrm{S}x), y(\in \mathrm{S}y)) \Leftrightarrow \in \mathbf{R}\mathrm{S} \qquad (4.15.1)\ \text{Df. } \mathbf{R}$$

$$\in \mathbf{S}, \mathrm{S}(\ddagger (\ ) \downarrow, x(\in \mathrm{S}x), y(\in \mathrm{S}y)) \Leftrightarrow \in \mathbf{Q}\mathrm{S} \qquad (4.15.2)\ \text{Df. } \mathbf{Q}$$

$$\in \mathbf{S}, \mathrm{S}('\in \mathbf{Q}\mathrm{S}, '\in \mathbf{R}\mathrm{S}) \Leftrightarrow \in \mathbf{M}\mathrm{S} \qquad (4.15.3)\ \text{Df. } \mathbf{M}$$

Perfect Sorts have all their members either identical or twins. The case with ordinable Sorts is tricky: if the definition given is held to be decidable, then by Axiom (4.13) it must follow that all the members of the Sort (for the predicate requires it to be a Sort) are bipar, from which it follows by (4.14) that it does not satisfy the definition of a Sort. Logically, according to (4.14), it still *might* be called a Sort, and in that case it is an ordinable Sort. But (4.15.2) also admits as ordinable any Sort for which the definition is undecidable in some cases, for then Axiom (4.13) does not apply; but this option will not arise in the present work. Ordinable Sorts are only a marginal, but nonetheless an important, case of Sorts; they have all the properties of Sets in regard to their own members, but differ from Sets in existing in an environment where twinship is not axiomatically excluded. Mixed Sorts are those where all three parity-relations are exemplified among their members.

I now define *elementary* Sorts **E** and *singular* Sorts **J**:

$$\in \mathbf{R}, \mathrm{S}(\in \mathrm{SS}) \Leftrightarrow \in \mathbf{E}\mathrm{S} \qquad (4.15.4)\ \text{Df. } \mathbf{E}$$

$$\in \mathbf{S}, \mathrm{S}(= (\ ) = , x(\in \mathrm{S}x), y(\in \mathrm{S}y)) \Leftrightarrow \in \mathbf{J}\mathrm{S} \qquad (4.15.5)\ \text{Df. } \mathbf{J}$$

We note that singular Sorts are both perfect and ordinable, and we can prove that

$$\in \mathbf{E}\mathrm{S} \Rightarrow \in \mathbf{J}\mathrm{S} \qquad (4.16)\ \text{Th.}$$

For if S has a member not identical with itself, this must either be distinct from S or twin to it. In the former case, there are distinct elements in S (for $\in \mathbf{E}\mathrm{S} \Rightarrow \in \mathrm{SS}$ by (4.15.4)) which contradicts the protasis of (4.15.4) which says S is perfect. In the latter case we have $\hat{=}\ \mathrm{S}s$ where s is the twin member, whence, applying (4.2.2) to the true relation $\in \mathrm{S}s$ we get $\in s\mathrm{S}$, and Axiom (4.1) then gives $= \mathrm{S}s$. Q.E.D. We can also prove that

$$\in \mathbf{E},x(\in,\mathrm{X}(\in \mathbf{J}\,\mathrm{X})x) \Rightarrow\ = \mathrm{X}x \qquad (4.17)\ \text{Th.}$$

that is, that any singular Sort, whose (sole) member is elementary, is identical with that member. Let us define / so that $/j$ is the Sort whose only members is j:

$$\in, \mathrm{J}(\in \mathbf{J}\,\mathrm{J})j \Rightarrow\ = /j\,\mathrm{J} \qquad (4.17\text{a})\ \text{Df.}/$$

Then $\in \mathbf{E}x$ from Theorem (4.16) implies $\in \mathbf{J}x$ that is $= /xx$, and so the X of (4.17) can be written $//x$. But if $= /xx$ then also $= //x/x$ and so $= //xx$ which is the same as $= \mathrm{X}x$, Q.E.D.

A Sort need not have any members; we can prove that

$$\in \mathbf{S}\mathrm{X},\quad {'\!\in} \mathrm{X}, u(\in \mathbf{S}u) \Rightarrow {'\!\in} \mathrm{X}, v(\ ) \qquad (4.18)\ \text{Th.}$$

That is, if X is Sort, of which all Sorts are not members, it has no members of any kind at all. Let us write $*\mathbf{S}$ for the class of all classes which are not Sorts, so that $\in *\mathbf{S}, x({'\!\in} \mathbf{S}x)$. Now if, as stated in the second protasis term, all Sorts are non-members of X (notice that $'\!\in$ is an ordinary relator, and is not of the form to which Axiom (4.13) takes exception), then any member of X must be a non-Sort, that is, $\in *\mathbf{S}, x(\in \mathrm{X}x)$. But, from (4.14), all pairs of members of a non-Sort are bipar, so that any two non-Sorts are distinct unless every member of each is a member of the other, in which case they are identical. Thus, any Sort having only non-Sorts as members is itself a non-Sort. Therefore, if X has any members, it must itself be a non-Sort, contradicting the first protasis term; it must therefore have none at all, Q.E.D.

Thus the members of a Sort, if any, must be Sorts. This is a further difference from Set theory, where Sets may have members which are not Sets. Thus the search for ultimate members of a Sort leads to an infinite recursion, which is the motive of defining elementary Sorts as being their own members. We may then define the *empty* Sort:

$${'\!\in}, \mathrm{Z}(\in \mathbf{S}\,\mathrm{Z}), x(\ ) \Rightarrow\ = \emptyset\,\mathrm{Z} \qquad (4.19)\ \text{Df.}\ \emptyset$$

If we now define the relator $\leqq$ by

$$\in, \mathrm{X}(\in \mathbf{S}\,\mathrm{X}), u(\in, \mathrm{Y}(\in \mathbf{S}\,\mathrm{Y})u) \Rightarrow\ \leqq \mathrm{X}\,\mathrm{Y} \qquad (4.20)\ \text{Df.}\ \leqq$$

we can prove the elementary theorem that

$$\in \mathbf{S}\mathrm{X} \Rightarrow\ \leqq \mathrm{X}\emptyset \qquad (4.21)\ \text{Th.}$$

For the relation $\in \mathrm{X}, u(\in \mathrm{Y}u)$ is false only if there is a counter-instance, i.e.

a member of Y such that $'\in Xu$. While ordinarily impossible to refute, hence Axiom (4.15), this is not so in the case where $= Y\emptyset$, for $\emptyset$ has no members and so there is no counter-instance, Q.E.D.

4.G. SOME CLASSIFYING FUNCTORS

There are two principal types of classifying functors used in **T**. One type has arguments which denote actual members of the Sort denoted, or in some cases Subsorts, and these I shall call *complexes*; the other type have a single argument incorporating a predicate defining what entities shall count as members of the Sort denoted, and are called *implexes*. The one classifying functor we have defined so far, the / of (4.17a), belongs to both types. In general, the problem in framing definitions of classifying functors is in the exclusion of unwanted members. For example, the statement

$$\in Z,x(=xy) \Rightarrow\; = Z/y \tag{4.22}$$

defines $/y$ as having y as one of its members, but if $\nmid xy$ the protasis, according to (4.12), leaves undecided whether $\in Zx$. However, this conflicts with the use of / in (4.17a); we have therefore to modify (4.22) by inserting an exclusion condition. We can write

$$\in Z,x(=x,z(\nmid z,y(\nmid yx))) \Rightarrow\; = /,x(=xz)Z \tag{4.23}$$

$$\Rightarrow\; = /zZ \tag{4.24}$$

Neither of these statements constitutes a definition sufficiently general for our requirements. They are equivalent to (4.17a) but without explicitly bringing in the concept of a singular Sort. To make a more generally useful functor out of this / we need to replace the z in the apodosis of (4.24) by a general predicate. (4.23) shows the form but of course as given it gives x only a single value. Taking due care (as in Set theory) to avoid situations liable to lead to antinomies, we write

$$\in \mathbf{S}\,D,\quad \leqq DZ,\quad \in \mathbf{P}\,D\,P,\quad \in Z,x(P,x(\nmid x,u('Pu))) \Rightarrow\; = Z/,x(Px)D \tag{4.25) Df. ,}$$

or in canonical form

$$(\ ) \Rightarrow {}^{+}\!\in [\,= Z/,x(Px)\,D\,]\,Z,x(,P(\in \mathbf{P},D(\in \mathbf{S},D(\leqq DZ))P),x(\nmid x,u('\,Pu))) \tag{4.25a}$$

Although the functor /, is defined with a second argument, specifying D the domain over which it can be trusted, I shall normally treat it as a univalent functor, the domain (as in analogous cases such as will be met with later) can ordinarly be assumed to be the most inclusive Sort currently under discussion. It will be seen that / by itself denotes the class of all possible values of its argument, and that the appended functor , serves as the usual means of indicating that there may be many of these.

Turning now to complexes, we first define the unordered pair functor ˆ as follows

$$\in \mathbf{S}Z,\quad \in Zx,\quad \in Zy,\quad {}'\!\in Z, u(\ddagger x, u(\ddagger uy)) \Rightarrow\, = Z\hat{\cup}xy \qquad (4.26)\ \text{Df.}\ \hat{}$$

or in canonical form

$$(\,) \Rightarrow {}^{*}\!\in[{}^{*}\!\in[{}^{*}\!\in[{}^{+}\!\in[\,= Z\,\hat{}xy]\,{}'\!\in Z, u(\ddagger x, u(\ddagger uy))]\, Z\hat{\cup}xy]\,\mathbf{S}Z \qquad (4.26a)$$

or

$$(\,) \Rightarrow\, = \hat{}xy, Z({}'\!\in, Z(\in \mathbf{S}, Z(\in Z\hat{\cup}xy)), u(\ddagger x, u(\ddagger x, u(\ddagger uy))) \qquad (4.26b)$$

Thus, $\hat{}xy$ denotes the Sort whose members are x, y and no others; one must check the context to ensure that x, y obey the concurrences needed to make this meaningful. As a rider, we can now define the class **V** of *unordered pairs* thus:

$$(\,) \Rightarrow\, = \mathbf{V}/\hat{}, x(\,), y(\,) \qquad (4.27)\ \text{Df. } \mathbf{V}$$

We can moreover generalize ˆ to a functor $\underset{n}{\hat{}}$ which I shall call the *multiplet* functor of *order n*. This is best done in Section 4K in terms of multiple pair-Sorts.

4.H. ORDERED PAIRS

All the classifying functors so far defined have for their values unordered Sorts, which may be perfect. There is also a class of functors which, applied to bipar arguments, produce ordered Sorts. The simplest of these is the *ordered pair* defined thus:

$$\in \mathbf{S}x,\quad \in \mathbf{S}y \Rightarrow\, = {}''xy\ \hat{}x\,\hat{}xy \qquad (4.28.1)\ \text{Df. }''$$

and the class of such pairs

$$\in \mathbf{Q}\,\mathrm{D} \Rightarrow\, = \mathbf{W}\,\mathrm{D}/'', x(\in \mathrm{D}x)\ , y(\in \mathrm{D}y) \qquad (4.28.2)\ \text{Df. } \mathbf{W}$$

Note that whereas the class **V** is restricted by no conditions, **W** is defined as having pairs of members of an ordinable Sort as its members. The functor $''$ however is not so restricted, though of course its value is different according as the arguments are or are not twins. If twins are in question, it matters that the x, y in the two terms of the equality on the apodosis are non-concurrent, on account of the classifying property of $\hat{}$; the consequence of this is that $''$ is defined only to within twinship, so that

$$\in \mathbf{S}v, \quad \in \mathbf{S}v, \quad \in \mathbf{S}w, \quad \in \mathbf{S}x, \qquad , \overset{!}{-}\, vx \;\Rightarrow \overset{!}{-}\, ''uv\,''wx \tag{4.29.1}$$

$$\overset{\wedge}{-}\, uw, \overset{\wedge}{-}\, vx \;\Rightarrow \top\, ''uv\,''wx \tag{4.29.2}$$

$$= uw, \; = vx = \Rightarrow \top\, ''uv, ''wx$$

(4.29.3) Th.

That is to say, ordered pairs whose members have twins are capable of only two parity-relations, distinct and indistinct; neither identity nor twinship can be meaningfully postulated as *different* relations between them. The following relations are easily verified:

$$'\in \mathbf{E}\,\hat{\cup}\, xy, \quad \overset{!}{-}\, xy \Rightarrow ''xy = \hat{}x\,\hat{}xy \ddagger \hat{}y\,\hat{}xy \tag{4.30.1}$$

$$\overset{\wedge}{-}\, xy \Rightarrow ''xy = \hat{}x\,\hat{}xy = \hat{}y\,\hat{}xy = \hat{}\hat{\cup}\, xy\,\hat{}xy \tag{4.30.2}$$

$$= xy \Rightarrow ''xy = \hat{}x\,\hat{}xx = \hat{}x/x \ddagger \hat{}xx \tag{4.30.3}$$

$$\in \mathbf{E}x \qquad \Rightarrow ''xx = \hat{}x\,\hat{}xx = \hat{}x/x = \hat{}xx = x$$

(4.30.4) Th.

The identity of $\hat{}x\,\hat{}xy$ and $\hat{}y\,\hat{}xy$ arises because since $\hat{}$ is classifying, the x outside can't be sure it isn't a y, and conversely.

The fact that the separated element in $\hat{}x\,\hat{}xy$ is in a different nexus from those in the last $\hat{}xy$ can also be expressed in the form

$$\in \mathbf{S}\,\hat{\cup}\, xy, \quad \overset{\wedge}{-}\, x\, y \Rightarrow = ''xy\,''yx \tag{4.31) Th.$$

from which follows the important corollary, that there is no ordering among twin elements of a Sort, so that the cardinal and ordinal of a Sort are in general different, the latter being the less except in ordinable Sorts, which are named for this property. All perfect Sorts have an ordinal of 1.

It is also occasionally useful to have functors inverse to $''$, denoting the first and second members of an ordered pair, defined thus:

$$\in \mathbf{W}p \Rightarrow = 1^{\circ} p, u(\in p, u(\in, q(\in pq)u)) \qquad \text{(4.32.1) Df. } 1^{\circ}$$

$$= 2^{\circ} p, u(= p\,''1^{\circ} pu) \qquad \text{(4.32.2) Df. } 2^{\circ}$$

These are of course undefined for unordered pairs.

4.J. SOME CONFOUNDING FUNCTORS

We have seen that we can define a functor $\hat{}$ which makes pairs out of its arguments; it is classifying by definition, and $\hat{}xy$ denotes the whole Sort, whose members are x, y, and not the members themselves. Closely related to $\hat{}$ there is another functor which denotes *not* the class whose members are x, y but any one of its members unspecified. We write $\hat{\smile}xy$ to denote an unspecified member of $\hat{}xy$, neither the Sort nor either of its particular members x or y. Its basic idea is that for any predicate P, $\mathrm{P}\hat{\smile}xy \Rightarrow \mathrm{P}x$; as usual, it takes longer to say this in properly rigorous form. The definition is:

$$\in\mathbf{S}\mathrm{D},\quad \in\mathbf{P}\mathrm{D}\mathrm{P},\quad \in\mathbf{P}\mathrm{D}\mathrm{Q},\quad \mathrm{P},u(,\mathrm{Q}(\mathrm{Q}x\,\mathrm{Q}y\,'\mathrm{Q},v(\text{ǂ}\,x,v(\text{ǂ}\,vy)))u) \Rightarrow \mathrm{P}\hat{\smile}xy \qquad \text{(4.33) Df. } \hat{\smile}$$

of which (4.26) is merely a special case where $= \mathrm{P}\in\hat{}xy$. By substituting x for u in the protasis, we find that Px is true, so that $\mathrm{P}\hat{\smile}xy \Rightarrow \mathrm{P}x$, as stated above, which is as defined for a confounding functor in 2.J. I shall call $\hat{\smile}xy$ the *blur* of x and y.

I have already on occasion used this functor in formulating canonical forms of protases, where its confounding property makes it useful because it facilitates concurrence; in any constraint $\in\mathrm{C}\hat{\smile}xy$, both x and y count as directly dependent on the governor $\in$ and so concurrent with any other x or y dependent on this $\in$. Another role of $\hat{\smile}$ is as the 'inverse' of /, as exhibited in the theorem

$$\in\mathbf{S}x, \in\mathbf{S}y \Rightarrow\ = \hat{}xy/\hat{\smile}xy \qquad \text{(4.34) Th.}$$

which follows directly from the definitions (4.25), (4.26), and (4.33). The pairing and classifying properties, combined in $\hat{}$, are separated between $\hat{\smile}$ and / in its equivalent $/\hat{\smile}$.

Closely related to $\hat{\smile}$ is the functor k defined as follows:

$$\in\mathbf{S}\mathrm{D},\quad \in\mathbf{P}\mathrm{D}\mathrm{P},\quad \mathrm{P},x(\in\mathrm{D}x) \Rightarrow \mathrm{Pk}\mathrm{D} \qquad \text{(4.35) Df. k}$$

It will be observed that $\hat{\smile}$ is equivalent to k $\hat{}$, and that the use of this functor obviates the need for such expressions as $\hat{\smile}\hat{\smile}\hat{\smile}wxyz$ which is more shortly expressed by $\mathrm{k}\hat{4}wxyz$. I shall call kD the *smudge* of the Sort D. Those elements of D which are not expressed as blurs but by single symbols are called *sharp* elements.

4.K. MISCELLANEOUS FUNCTORS OVER SORTS

I conclude this chapter by setting out the definitions of a variety of functors which have for the most part obvious analogues in Set theory, but whose definitions have to be reconsidered in Sort theory so as to accord, where possible, with the constraints of the notation. First, there is one univalent functor, the *power-Sort*, which I write $!2$. If this is defined in the obvious way, as $/,u(\leqq \mathbf{S}u)$, we have $\in !2\mathrm{X}\,\mathrm{X}$ thereby making $!2$ a classifying functor of X. It does not seem worth while to avoid this, but it is important to remember that, unlike those which follow, $!2$ is formally classifying. Thus I define

$$\in \mathbf{S}\,\mathrm{X} \Rightarrow = !2\mathrm{X}\,/,u(\leqq \mathrm{X}u) \qquad \text{(4.36) Df. } !2$$

There is in **T** no absolute complement of a Sort, but only relative complements, or *Sort Remainders* defined as follows:

$$\in \mathbf{S}\,\hat{\cup}\,\mathrm{XY} \Rightarrow = \text{`}\mathrm{XY}\,/,u('\in \mathrm{Y},u(\in \mathrm{X}u)) \qquad \text{(4.37) Df.}$$

Another relation which escapes us in a general form, but not in all cases, is the union of two Sorts. The intersection however can be defined quite generally, thus:

$$\in \mathbf{S}\,\hat{\cup}\,\mathrm{XY} \Rightarrow \wedge\mathrm{XY}\,/,u(\in \mathrm{X},u(\in \mathrm{Y}u)) \qquad \text{(4.38) Df. } \wedge$$

and we can then define the union of any Subsorts *of a given* D by

$$\in \mathbf{S}\,\mathrm{D}, \quad \leqq \mathrm{D}\,\hat{\cup}\,\mathrm{XY} \Rightarrow = \cup\mathrm{XY}\ \text{`}\mathrm{D} \wedge \text{`}\mathrm{DX}\ \text{`}\mathrm{DY} \qquad \text{(4.39) Df. } \cup$$

Any attempt to define $\cup$ on the model of (4.38) involves negating an expression containing , contrary to Axiom (4.13).

A very useful functor is the unordered *pair-Sort* namely

$$\in \mathbf{S}\,\hat{\cup}\,\mathrm{XY} \Rightarrow = \hat{\wedge}\mathrm{XY}\,/,p(\ddagger\,\emptyset \wedge \mathrm{X},p(\ddagger\,\emptyset \wedge \mathrm{Y},p(\in \mathbf{V}p))) \qquad \text{(4.40) Df. } \hat{\wedge}$$

which makes the pair-Sort of X and Y the Sort of all unordered pairs which have a member in each of X, Y. Generalizing, we have also the unordered *Sort-power* defined by

$$\in \mathbf{S}\mathrm{X}, \quad \in \mathbf{I}n \Rightarrow = \hat{.}\mathrm{X}n\ \hat{\wedge}\mathrm{X}\,\hat{.}\mathrm{X}(n-1), \quad = \hat{.}\mathrm{X}1\,\mathrm{X} \qquad \text{(4.41) Df. } \hat{.}$$

and the *multiplet* of order n over X, which is defined by

$$\in \mathbf{S}\mathrm{X}, \quad \in \mathrm{X}x_{,\,i(\in \mathbf{I}\,i)} \Rightarrow = \hat{n}\,x_1 \ldots x_n,u(\in \hat{.}\mathrm{X}nu) \qquad \text{(4.42) Df. } \hat{n}$$

There are also, in appropriate cases, *ordered* pair-Sorts, Sort-powers, and multiplets. Definitions of the last two differ from their unordered correspondents only by replacing the functors ˆ̦, ˆ by ⁞, '' ; but the ordered pair-Sort needs a special definition:

∈S ˆ̮XY ⇒ = ⁞XY /, p(∈X1⁰, p(∈Y2°, p(∈**W**p))) (4.43) Df. ⁞

I have expressed the above definitions, for the sake of clarity, using the classifying functor /. As a result, ambiguities can arise in the interpretation of the arguments X, Y, and I must substantiate the definitions by giving their canonical forms. These are:

(4.36) *∈[⁺∈[= '.2X Z]∈, Z('∈Z, U(' ≦ XU)), V(≦ XV)]**S**X (4.36a)'.2

(4.37) *∈[⁺∈[= `XY Z]∈, Z('∈Z, u('∈Xu)), v(∈X, v('∈Yv))]**S** ˆ̮XY (4.37a) `

(4.38) *∈[⁺∈[= ˬXY Z]∈, Z('∈Z, u('∈Xu), '∈Z, v('Yv)), w(∈X, w(∈Yw))]**S** ˆ̮XY (4.38a) ˬ

(4.39) already in canonical form

(4.40) *∈[⁺∈[= ˆ̦XY Z]∈, Z('∈Z, q(∉∅ `**S** ˬXYq), '∈Z, q(∈ `**SV**q)), p(∉∅ ˬX, p(∉∅ ˬY, p(∈**V**p)))]**S** ˆ̮XY (4.40a) ˆ̦

(4.41) already in canonical form
(4.42) already in canonical form

(4.43) *∈[⁺∈[= ⁞XY Z]∈, Z('∈Z, q(∉∅ `**S** ˬXYq), '∈Z, q(∈ `**SW**q)), p(∈X1⁰, p(∈**V**2⁰, p(∈**W**p)))]**S** ˆ̮XY (4.43a) ⁞

CHAPTER V

SORT THEORY – MAPPINGS

Having in the last chapter explained the basic concepts of Sort theory, and proved a few mostly rather trivial theorems, I now return to consider the questions raised in Chapter III, namely how to construct valid representations of Sorts. Obviously, any such representation will be some kind of mapping; equally obviously, the concept of a 'mapping' is likely to appear in Sort theory, if at all, in a more or less distorted form. In this chapter, I shall try to show that the necessary logical properties which attach to the main types of mappings in Set theory do apply also to certain mappings between Sorts, and that by their means we can apply the ideas of Chapter III to Sorts. The mathematical procedures required will be explained and applied in the course of Chapters VI and VII.

5.A. MAPPINGS AND FUNCTIONS

We are accustomed in Set theory, and in general mathematics, to identify 'functions' with 'mappings'. Both are described as sets of ordered pairs, of which one element comes from the domain and the other from the range of the mapping, the argument and the value of the function respectively. A functor, however, denotes not a set of ordered pairs, but an operation whereby one element of such a pair is replaced by the other. It is of course assumed that, given the set of pairs, it will be possible to pick out the one required, identify its elements, and perform the substitution, in any and every case where a functor appears.

In a triparitous context, as will by now be clear, none of this can be taken for granted; nor need there be a one–one relationship between functors and mappings. The easier of the two concepts to expound is that of a mapping; but functors are easier to use. To begin with we would like to define a functor specifying that given members of a Sort exhaust its membership; this, though obscurely, can be stated thus:

$$\in \mathbf{S}\mathrm{Z},\quad \in \mathbf{P}\mathrm{P},\quad \in_{,} \mathrm{Z}('\in \mathrm{Z}_{,} u('\mathrm{P}u)),_{,} v(\mathrm{P}v) \Rightarrow \hat{\in} \mathrm{Z}_{,} x(\mathrm{P}x) \qquad \text{(5.1) Df. } \hat{\in}$$

and with its help we can define a *mapping* from a domain D to a range R by:

$$\in \mathbf{S}\,\hat{\cup}\,\mathrm{RD},\quad \in \mathbf{W}p,\quad \in \mathbf{F}_{,} p(\in \mathrm{R}2^{0}{}_{,} p(\hat{\in} \mathrm{D}1^{0} p)) \Rightarrow \in^{!}\mathrm{RD}\mathbf{F}$$

(5.2) Df. $^{!}$

so that **F** is a Sort of ordered pairs, whose first members exhaust the domain. To pass from mappings to functors, we define

$$\in \mathbf{S}\hat{\cup}\mathrm{RD}, \quad \in\!{}^{!}\mathrm{RDF}, \Rightarrow = |\mathbf{F}, x(= x1^{0}, p(= 2^{0}, p(\in \mathbf{F}p)y))y$$

(5.3) Df. |

where the nonce-symbol | is only there so to have something to define for I shall treat the subformula |**F** as equivalent to F. This is because we shall use functors F much more often than the mappings F they refer back to.

5.B. MAPPINGS FROM AND TO PERFECT SORTS

A mapping **F** is a class of ordered pairs, and so the only difference between a mapping belonging to ${}^{!}$NM and one belonging to ${}^{!}$MN is the ordering of the pairs. By contrast, the corresponding functors behave very differently. Let us suppose that = M $\hat{3}qrs$ and = N $\hat{\ }uv$, and consider what difference it makes if either is a perfect Sort. If M is perfect, and the mapping **F** is formally defined by = **F** $\hat{3}$ ''*qu*''*rv*''*su*, then if N is ordinable the three relations using the functor F namely = F*qu*, = F*rv*, = F*su*, can be taken more or less at face value; that is to say, we cannot state that = F*q* $\hat{\cup}$ *uv*, because since ${}^{!}uv$ $\nmid$ F*qv* (more precisely, $\nmid$ F, *q*(= F*qu*)*v*). We *can* say = F $\hat{\cup}$ *qsu* of course, but we still know that *q*, *r*, *s* are three entities not one, and that = F*qu* is true of one of them but not of all. If, on the contrary, M is ordinable and N perfect, the same specification of **F** leads us to infer from = F*qu* that equally correctly (from (4.2.2)) we can say = F*qv*, i.e. ( ) ⇒ F*q* $\hat{\cup}$ *uv*. We know that there are two possible images for each element of M, but we cannot know which in any given case, whether F*q* or any other. But we can state that = F*q*F*s* and $\nmid$ F*q*F*r*, because in an ordinable M *q*, *r* have no twins and so their non-concurrence in these expressions is nugatory.

These results, though elementary, are important, and I shall state them here as a theorem:

$$\in \mathbf{Q}\mathrm{Q}, \quad \in \mathbf{R}\mathrm{R}, \quad \in\!{}^{!}\mathrm{QR}\,\mathbf{F} \Rightarrow \nmid \mathrm{F}, q(\in \mathrm{R}q)\,\hat{\cup}x, y(\nmid xy) \qquad (5.4.1)$$

$$'\!\in \mathbf{2} = \mathrm{F}q\,\mathrm{F}r \qquad (5.4.2)$$

$$\in\!{}^{!}\mathrm{RQ}\,\mathbf{E} \Rightarrow = \mathrm{E}, y(\in \mathrm{Q}y)\,\mathrm{kR} \qquad (5.4.3)$$

$$= (\ ) = \mathrm{E}x\mathrm{E}y \qquad (5.4.4)$$

The apodosis of (5.4.1) states that the value of any F*q* is a sharp element (of the range), and not a blur; this contrasts with (5.4.3) which says that if the range is perfect, the value of any Ey is the smudge of the whole range. (Where the range is mixed, that is neither perfect nor ordinable, the value of a formula such as Fx is in general a blur, but not the smudge of the whole range.) Note, however, that this is not because distinct entities do not have blurs; the concept of having something which might be one of two distinct things, such as a black ball or a white one, is well-known in statistics, for example, even if the blur notation is not employed. The results (5.4) are specific consequences of translating the concept of a mapping into a triparitous theory.

Parity-relations between functors defined over a given domain can be defined as follows:

$$\begin{aligned} \in\!\overset{!}{.}\mathrm{RD}\,\hat{\cup}\,\mathbf{EF} \;=\; & \emptyset/,x(\nmid \mathrm{E}x\mathrm{F}x) && \Rightarrow\; = \mathrm{EF} \\ \nmid\; & \emptyset/,x(\nmid \mathrm{E}x\mathrm{F}x), \quad = \emptyset/,x(\overset{!}{\_}\mathrm{E}x\mathrm{F}x) && \Rightarrow \overset{\wedge}{\_}\;\mathrm{EF} \\ & \qquad\qquad\qquad \nmid\, \emptyset/,x(\overset{!}{\_}\mathrm{E}x\mathrm{F}x) && \Rightarrow \overset{!}{\_}\;\mathrm{EF} \end{aligned} \qquad (5.4a)$$

Unlike mappings, defined as classes of ordered pairs, functions, which are denoted by functors, are not classes, and have no members except themselves (as follows from Theorem (4.17), though unlike identifiers they do generate classes, in the sense exemplified in (5.4)).

5.C. THE CLOSURE OF A SORT

It is evident from the foregoing discussion that the image of any element of a domain under a mapping onto a perfect range is in general a smudge. This makes it necessary to include at least some blurs of range elements among the effective elements of the range. This matters chiefly in cases where the domain and the range of a mapping are identical (where conventionally the term domain is used for both); for if blurs are excluded, we shall not find that any domain is *closed* under a mapping, which we would wish to be able to say this of. I therefore propose the following definition:

$$\in \mathbf{S}\mathrm{Z}, \quad \leqq \mathrm{ZD}, \quad \hat{\in}\mathrm{Z}\,\hat{\cup}\,,x(\in \mathrm{D}x), y(\in \mathrm{Z}y) \Rightarrow\; = \bar{\mathrm{D}}\,\mathrm{Z} \qquad (5.5)\ \mathrm{Df.}^{-}$$

The expression $\bar{\mathrm{D}}$ is said to denote the *closure* of D. A univalent functor is further said to be closed over D iff $\in \bar{\mathrm{D}}\,\mathrm{F}_{,}\,x(\in \bar{\mathrm{D}}x)$, and a bivalent functor G is so described iff $\in\bar{\mathrm{D}}\mathrm{G}\,{}_{,}p(\in\hat{:}\bar{\mathrm{D}}2p)$; conversely, $\bar{\mathrm{D}}$ may be said to be closed *under* any such functor F or G.

It may be noted that the elements of $\bar{\mathrm{D}}$ can be regarded as the smudges of the Subsorts of D, the smudge of a singular Subsort being identical with its only member. Thus $\bar{\mathrm{D}}$ has the same cardinal as $\dot{!}2\mathrm{D}$ but it has a very different structure. The power-Sort of a perfect Sort of cardinal n is a mixed Sort, with cardinal 2^n and ordinal $n + 1$ (both diminished by 1 if the empty Sort is excluded). The closure is a perfect Sort, of cardinal $2^n - 1$ and ordinal 1. To prove this we shall have to examine the properties of the functor $\hat{\cup}$. First, I shall prove that $\hat{\cup}$ is associative:

$$\in\mathbf{S}x,\quad \in\mathbf{S}y,\quad \in\mathbf{S}z \Rightarrow\, = \hat{\cup}\hat{\cup}xyz\,\hat{\cup}x\hat{\cup}yz \qquad \text{(5.6.1) Th.}$$

For, from (4.33) where $\hat{\cup}$ is defined, the formula $\hat{\cup}\hat{\cup}xyz$ can be expressed in the form $,v('\mathrm{Q},u(\nmid x,u(\nmid y,u(\nmid uz))),\mathrm{Q}(\mathrm{Q}x\mathrm{Q}y\mathrm{Q}z)v)$; and since the order of nested ,-expressions is commutative (see 4.E) as is of course that of the terms of the conjunction $(\mathrm{Q}x\mathrm{Q}y\mathrm{Q}z)$, the same expression is obtained for the formula $\hat{\cup}x\hat{\cup}yz$, which therefore equals the other, Q.E.D. Likewise it follows that $\hat{\cup}$ is idempotent:

$$\in\mathbf{S}x \Rightarrow\, = \hat{\cup}xxx \qquad \text{(5.6.2) Th.}$$

and from these two together that

$$\in\mathbf{S}x,\quad \in\mathbf{S}y \Rightarrow\, = \hat{\cup}xy\,\hat{\cup}x\hat{\cup}xy \qquad \text{(5.6.3) Th.}$$

For from (5.6.1) $\hat{\cup}x\hat{\cup}xy$ is identical with $\hat{\cup}\hat{\cup}xxy$ which equals $\hat{\cup}xy$, Q.E.D. We are now in a position to prove that the closure of a perfect Sort is perfect. First, I prove that

$$\in\mathbf{S}x,\quad \in\mathbf{S}y,\quad \hat{-}xy \Rightarrow \nmid x\hat{\cup}xy \qquad \text{(5.7) Th.}$$

Consider the ordered pairs $''x\hat{\cup}xy$ and $''\hat{\cup}xyx$: the first reduces by (4.28) to $\hat{\ }x\hat{\ }x\hat{\cup}xy$, where the x's are non-concurrent because $\hat{\ }$ is classifying. So, from (4.2.2) we can equate this to $\hat{\ }y\hat{\ }x\hat{\cup}xy$ which along with the previous formula gives us $\hat{\ }\hat{\cup}xy\hat{\ }x\hat{\cup}xy$ which, by (4.24) again, identifies with $''\hat{\cup}xyx$. Therefore these pairs are not in fact ordered, and so their terms are not distinct, Q.E.D.

It is now a simple corollary of Theorem (5.7), following by iteration, that all blurs over a perfect Sort are twin to each other and to the sharp elements; that is to say

$$\in\mathbf{R}\mathrm{D} \Rightarrow \in\mathbf{R}\bar{\mathrm{D}} \qquad \text{(5.8) Th.}$$

This has the consequence that though the 'sharp' elements of D form a Subsort of its closure $\bar{\text{D}}$, it makes no sense to identify any one member of $\bar{\text{D}}$ as being sharp, and others blurs. This fact imposes an important constraint on what representations of a Sort can be accepted as valid, which we shall encounter later, in 6.E, 6.H, 6.J.

We can also prove that all functors are distributive over $\hat{\cup}$:

$$\in \mathbf{S}\,\mathrm{D},\quad \in \mathrm{D}\,\hat{\cup}\,\hat{\cup}\,xyf\,\hat{\cup}\,xy \Rightarrow\; = \hat{\cup}\,fxfyf\,\hat{\cup}\,xy \qquad \text{(5.9) Th.}$$

For let P be any predicate; then, from (4.33), $\mathrm{P}\,\hat{\cup}\,fx\,fy$ is the disjunction of $\mathrm{P}fx$ and $\mathrm{P}fy$. If now B is a relator and a is a constant, $\mathrm{B}ax$ is a predication on an argument x. We may therefore compare $\mathrm{P}fx$ with the formula $\mathrm{B}afx$. It follows from Theorem (2.12) if f is non-classifying, and from Theorem (2.6a) if it is classifying, that $\mathrm{P}f$ is also a predicate. Thus, from (4.33) again, $\mathrm{P}f\,\hat{\cup}\,xy$ is the disjunction of $\mathrm{P}fx$ and $\mathrm{P}fy$; therefore $\mathrm{P}\,\hat{\cup}\,fxfy$ is equivalent to $\mathrm{P}f\,\hat{\cup}\,xy$ and since P is an unspecified predicate, (5.9) follows.

5.D. A CLASSIFICATION OF SORT MAPPINGS

It will by now be evident that some types of mappings between Sorts are highly degenerate, while others are not. It is useful at this point to give a brief classification and define some terms:

$$\in^{!}\mathrm{RD}\mathbf{F},\ \in \mathbf{Q}\mathrm{R},\ \nmid \mathrm{F},x(\hat{-}x,y(=\mathrm{F}ya))a \Rightarrow \mathbf{F}\text{ is a }\textit{Setlike}\text{ mapping} \qquad (1)$$

$$\in \mathbf{R}\,\mathrm{R},\ \nmid \mathrm{D}\,\hat{:}\,\mathrm{R}n,\ = \emptyset \wedge \mathrm{DR} \Rightarrow \mathbf{F}\text{ is a }\textit{Smudged}\text{ mapping} \qquad (2)$$

$$= \mathrm{D\,R} \Rightarrow \mathbf{F}\text{ is an }\textit{Endomorphism} \qquad (3)$$

$$= \mathrm{D}\,\hat{:}\,\mathrm{R}n \Rightarrow \mathbf{F}\text{ is a }\textit{Function}\text{ of }n\text{ arguments} \qquad \text{(5.10.4) Dff.}$$

There are obviously many possibilities not named here, whose study may well prove rewarding to the mathematician. I shall here be concerned mainly with functions of two arguments, and with 'endomorphisms', a term whose meaning here has been somewhat widened from its usual biparitous connotation.

Setlike mappings have the property that, even if the domain D is a perfect Sort, the mapping itself is ordinable, as is its range, and they have the same cardinal. This fact will be exploited in the next section to define the cardinal of a perfect Sort. One familiar property which such mappings have only if the domain also is ordinable, is that of being invertible. In other respects, they behave much like mappings between Sets. Smudged mappings are the most trivial. They map every element of the domain onto the smudge of the range Theorem (5.4.3). Endormorphisms exhibit an intermediate degree of degeneracy, and will call for further study.

5.E. CARDINALITY OF PERFECT SORTS

The problem of finding the cardinal of a perfect Sort is that there is no analogue of counting; one cannot know, among indistinguishables, which have been counted and which remain to count. It does appear, however, that a perfect Sort can be mapped onto an ordinable one, say the Integers, and that such a mapping could provide a solution to the problem.

For completeness' sake we begin by defining, within the notation $\mathscr{T}$, the cardinality of an ordinable Sort; this requires first a definition of the integers, via the relator $>$, of the functor $+1$, and then of a functor ∂ defined for ordinable Sorts to denote counting, and finally of the integer 0. Thus:

$$\in \mathbf{C}\hat{\cup}\hat{\cup}xyz,\ {=}(/)\mathrm{R}xx,\ {=}(/)(\mathrm{R}xy\mathrm{R}yx),\ {=}(/)(\mathrm{R}xy\mathrm{R}yx'\mathrm{R}xz) \Rightarrow {=}>\mathrm{R} \qquad (5.11.1)\ \text{Df.} >$$

$$\in \mathbf{C}\hat{\cup}\hat{\cup}xy\mathrm{F}\hat{\cup}xy,\ >x\mathrm{F}x,\ {=}xy,\ {=}\mathrm{F}x\mathrm{F}y \qquad \Rightarrow {=}+1\mathrm{F} \qquad (5.11.2)\ \text{Df.}+1$$

$$\in \mathbf{C}\mathrm{Z},\quad \in \mathrm{Z}{,}x(>{,}y(\in \mathrm{Z}y)x),\quad \nmid \mathrm{Z}\emptyset \qquad \Rightarrow {=}\mathbf{I}\,\mathbf{Z} \qquad (5.11.3)\ \text{Df. } \mathbf{I}$$

$$\in \mathbf{Q}\mathrm{U} \Rightarrow {=}\partial\mathrm{U}\ `\mathrm{U}{,}\mathrm{J}(\leqq \mathrm{U}{,}\mathrm{J}(\in \mathbf{J}\,\mathrm{J})) \qquad (5.11.4)\ \text{Df. } \partial$$

$$\in \mathbf{Q}\hat{\cup}\mathrm{UFU},\quad \in^{!}\mathbf{IS}\,\mathbf{F},\quad {=}\mathrm{FU}+1\mathrm{F}\partial\mathrm{U} \qquad \Rightarrow {=}{}^{c}\mathrm{F} \qquad (5.11.5)\ \text{Df. }{}^{c}$$

$$(\ ) \qquad \Rightarrow {=}0\,{}^{c}\emptyset \qquad (5.11.6)\ \text{Df. } 0$$

The relator $>$ generates an ordering of classes; the functor $+1$ denotes a class with 'one' member more than its argument has, and **I** is a class whose members are ordered by $>$ and therefore by $+1$. It is not necessary to specify, as in a biparitous theory, any parity-relation between classes with the same number of different members; they will be understood to be twins, and thereafter they will be tacitly confounded simply by ignoring the concurrence rules. There is of course a question how the functor ∂, which applied to an ordinable Sort removes one of its elements, will stand up when we come to apply the definition of ^{c} to perfect Sorts; for I have already argued that the counting procedure which underlies the definition (5.11.4) does not work in this case. We can however evade this question by using the definition:

$$\in \mathbf{R}\,\mathrm{U} \Rightarrow {=}{}^{c}\mathrm{U}\,{,}x(\leqq{,}x(\geqq{}^{c}{,}\mathbf{F}(\in^{!}\mathbf{I}\mathrm{U}\,\mathbf{F})\,x)x) \qquad (5.12)\ \text{Df. }{}^{c}$$

As we have seen, any mapping from a perfect to an ordinable Sort is itself an ordinable Sort, and so has a cardinal defined by (5.11.5). Then x is required to be greater than or equal to the cardinal of *any* such mapping, that is, it can have any value not less than the largest such cardinal; and the same x is also required to be *less* than or equal to any such value. To satisfy both conditions it must be equal to the cardinal of the largest mapping in $\overset{!}{.}$IU. There are such mappings where every member of U maps onto a different integer; none can have a larger cardinal than one of these, though mappings which carry two or more elements of U onto one integer are smaller. Therefore the integer specified by (5.12) as the cardinality of a perfect Sort must be the number of its (non-identical) members. And it has been found without applying ∂ to the perfect Sort U. The same definition (5.12) can be applied to ordinable and mixed Sorts also, though not without redundancy in the former case.

It therefore follows that the definition (5.12) can be used to define the cardinal of any Sort, though of (5.11) the only definition which we have not used is that of ∂ (and 0, which will remain useful). The existence of such cardinals is perhaps intuitively obvious from the original characterization of indistinguishables in (1.1): but we have come a long way from this beginning, and it is satisfactory to see that the theory still holds together. Since the cardinals are all distinct from each other, it follows at once that

$$\in \mathbf{S}\hat{\cup}XY, \quad \ddagger\, {}^{c}X\,{}^{c}Y \Rightarrow \overset{!}{-} X\,Y \qquad \text{(5.14) Th.}$$

for if not, i.e. if $\dagger XY$, we have from (4.2.2) that $= \leqq XY \leqq YX$, which can only be true if the cardinals are equal.

We can use the above result to set out the conditions for Sorts to satisfy the various parity-relations. This set of statements takes the place in Sort theory of the so-called Axiom of Extent in Set theory. But in the present context, since they are all inferrable from what has gone before, their status is not that of axioms but theorems. This difference seems to depend on the absence, in my theory, of the Set – Member distinction. These theorems are:

$$\in \mathbf{S}\hat{\cup}XY, \quad = {}^{c}X\,{}^{c}Y, \quad = {}^{c}Y\,{}^{c}/,'' xx(\in \ddot{\shortmid} XY\,'' xx) \Leftrightarrow = X\,Y \qquad \text{(5.15.1)}$$

$$< {}^{c}Y\,{}^{c}/,'' xx(\in \ddot{\shortmid} XY\,'' xx) \Leftrightarrow \overset{\wedge}{-} X\,Y \qquad \text{(5.15.2)}$$

$$\ddagger\, X\,Y, \quad \hat{+} X\,Y \qquad \Leftrightarrow \overset{!}{-} X\,Y \qquad \text{(5.15.3) Th.}$$

The first two of these depend on the enumeration of the identical pairs in an ordered pair-Sort, which equal in number the members of the parent

Sorts only if the latter are identical; this evades the ambiguities in the Set theory concept, of sharing identical members, which arise when it is transposed into a triparitous context.

5.F. THE INVARIANT SUBDOMAIN THEOREM

I now come to consider the two most important classes of Sort mappings enumerated in (5.10), namely 'endomorphisms' and functions. By an endomorphism over a Sort domain D we mean (following (5.10.3)) any mapping from the closure $\bar{\mathrm{D}}$ onto itself; alternatively, if **E** is an endomorphism defined over D, this can be expressed in the form $\in \bar{\mathrm{D}}\mathrm{E}, x(\in \mathrm{D}x)$, as will be proved below.

The fundamental theorem concerning endomorphisms, which I call the 'invariant subdomain theorem', may be stated as follows:

$$\in \dot{!}\,\bar{\mathrm{D}},\mathrm{D}(\in \mathbf{R}\,\mathrm{D})\,\hat{\cup}\,\mathbf{F},\mathbf{G}(=/,u(=\mathrm{F}uu)/,v(=\mathrm{G}vv)) \Leftrightarrow =\mathrm{F}\,\mathrm{G} \qquad (5.16)\ \mathrm{Th.}$$

That is to say, if either **F**, or any **G** having an identical Subdomain of D invariant under it, is an endomorphism mapping D into its closure $\bar{\mathrm{D}}$, then the functors F, G are identical; the point being that they will not be distinguished by the fact that there is some x such that $=\mathrm{F}xy$, $=\mathrm{G}xz$, $\nmid yz$, which one might think could happen. It is obvious that if the invariant subdomains differ, then $\nmid \mathrm{F}\,\mathrm{G}$; surprisingly, the converse is also true. Suppose that there is some x such that $\nmid \mathrm{F}x\,\mathrm{G}x$; since the two x's in this expression are non-concurrent, it is simply equivalent to $\nmid \mathrm{F}xy$, and we want therefore to assert or deny, unambiguously, this expression for particular non-identical elements x, y. Let us consider in what ways this can be stated in **T**: (1) $\nmid \mathrm{F}xy$: if Fx is known, y is something else, and in the absence of further particulars we may conclude $=\mathrm{F}xk$ where $=kk\mathrm{D}$; (2) introducing a condition, try $\nmid \mathrm{F}x\,,y(\nmid xy)$: if y is not x, 'Fx' is not y, but as the x's are non-concurrent this could mean anything; (3) placing the same condition on x, we get $\nmid \mathrm{F}\,,x(\nmid xy)\,y$: if x is not y, Fx is not y, and as before Fx may or may not be x, and again $=\mathrm{F}xk$; (4) we can, however, pre-select x, y as elements of D, in the form $\in \bar{\mathrm{D}}\,\hat{\cup}\,xy$, $\nmid \mathrm{F}xy$: but this in fact says nothing other than (1) above; (5) so what about $\in \bar{\mathrm{D}}\,\hat{\cup}\,xy$, $\in \bar{\mathrm{D}}\,\hat{\cup}\,x,y(\nmid xy)$, $\nmid \mathrm{F}xy$: this fails because the final y is non-concurrent with the first one, and says nothing; finally (6) $\in \bar{\mathrm{D}}\,\hat{\cup}\,xy$, $\nmid \mathrm{F}x,y(\nmid xy)$ *first* selects the elements x, y in the blur $\hat{\cup}\,xy$, and afterwards imposes the condition on them $\nmid xy$, which is illegitimate in view of rule

V1.1 in Section 1.7, unless we accept that it is denying the same thing as (5), which has already shown to be vacuous. Therefore the only meaning we can attach to the expression ⫲ FxGx is that either Fx or Gx is equal to the smudge of the domain. Therefore if ⫲ F G, they must have different invariant subdomains, Q.E.D.

We have therefore as a corollary of the Theorem (5.16) that

$$\in\dot{!}\,\bar{\mathrm{D}},\mathrm{D}\,(\in \mathbf{R}\,\mathrm{D})\mathbf{F} \Rightarrow\ = \mathrm{F},x('\in/,u(=\mathrm{F}uu)x)\mathrm{kD} \qquad (5.17)\ \text{Th.}$$

It follows from this that the number of non-identical endomorphisms F on a given domain is equal to the number of non-identical subdomains. All of these endomorphisms are mutually twins, for to be distinct there must be at least one case where $\dot{-}$ Fx Gx, and since they are endomorphisms both Fx and Gx are in D̄ and therefore $\hat{-}$ Fx Gx. The Sort of endomorphisms on a given D is therefore perfect. Its cardinal however is not equal to $\dot{!}$2 D, the power-Sort. For if the invariant subdomain of F is empty, = F, x(∈Dx)k, so **F** is formally a 'smudged mapping', and the relation between its domain and its range is indeterminable, so that its status as an endormorphism rests not on its performance, but solely on its definition. I therefore exclude this mapping, which has no invariants, from the class of true endomorphisms over a given domain.

This can be formalized in the definition

$$\in \mathbf{R}\,\mathrm{D},\quad \in\dot{!}\,\bar{\mathrm{D}}\mathrm{D}\,\mathbf{E},\quad ⫲\,\emptyset/,u(=\mathrm{E}uu) \Rightarrow \in \mathbf{En}\,\mathrm{D}\,\mathrm{E} \qquad (5.18.1)\ \text{Df. } \mathbf{En}$$

The concurrence of the D in the apodosis with that in the protasis is validated by the canonical form

$$*\in[*\in[\,⫲\,\emptyset/,u(=\mathrm{E}uu)]\,\dot{!}\,\bar{\mathrm{D}}\mathrm{D}\,\mathrm{E}]\,\mathbf{R}\,\mathrm{D} \Rightarrow \in \mathbf{En}\,\mathrm{D}\,\mathrm{E} \qquad (5.18.1a)$$

It has the consequence that the endomorphisms over D number 2^cD – 1

$$\in \mathbf{R}\,\mathrm{D} \Rightarrow\ = \dot{!}'\,2^c\mathrm{D} + 1^c\mathbf{En}\,\mathrm{D} \qquad (5.18.2)\ \text{Th.}$$

There is a large field of research on endomorphisms over mixed Sorts, which I do not follow up here, though there might well be intresting results to be obtained.

5.G. FUNCTIONS OF TWO ARGUMENTS OVER A PERFECT SORT

There are two Sorts of two-argument functions over a perfect domain D, namely $\dot{!}$ D̄ $\hat{\dot{}}$ DD and $\dot{!}$' D̄D $\hat{\dot{}}$ DD (note that $\hat{\dot{}}$ is non-classifying, despite its relation to ˆ); I deal with the former first. The general form of members of this Sort is '' ˆxyz, where x, y, z are all indistinct; the ordered pair is of course in order, since in general $\underset{-}{!}\hat{}xyz$ by Theorem (5.14). From Theorem (4.30) the number of distinct values of this formula depends on whether the elements are themselves elementary Sorts or not. I therefore distinguish between *prime* Sorts $\mathbf{R}_0$ and *derived* Sorts $\mathbf{R}_1$, defined thus:

$$\in \mathbf{R}\,\mathrm{D}, \quad \in \mathbf{E}\,{}_{,}x(\in \mathrm{D}x) \Leftrightarrow \in \mathbf{R}_0\,\mathrm{D} \qquad (5.19.1)\ \text{Df. } \mathbf{R}_0$$

$$'\!\in \mathbf{E}\,{}_{,}x(\in \mathrm{D}x) \Leftrightarrow \in \mathbf{R}_1\,\mathrm{D} \qquad (5.19.2)\ \text{Df. } \mathbf{R}_1$$

(Note that no *perfect* Sort can have elements of both kinds). We can now classify the possible values of $''\,\hat{}xyz$ according to the degree of degeneracy involved into four Subsorts of a given **F**, thus:

Subsort:	$\mathbf{F}_0$	$\mathbf{F}_1$	$\mathbf{F}_2$	$\mathbf{F}_3$	
containing pairs of the form:	$''\,\hat{}xyz$	$''\,\hat{}xyy$	$''\,\hat{}xxy$	$''\,\hat{}xxx$	
which reduce to					
if $\in \mathbf{R}_0\,\mathrm{D}$:	$\hat{}\hat{}xy\ \hat{}\hat{}xyz$	$\hat{}\hat{}xy\ \hat{}\hat{}xyz$	$\hat{}x\ \hat{}xy$	x	
if $\in \mathbf{R}_1\,\mathrm{D}$:	$\hat{}\hat{}xy\ \hat{}\hat{}xyz$	$\hat{}\hat{}xy\ \hat{}\hat{}xyz$	$\hat{}/x\ \hat{}/xy$	$\hat{}/x\ \hat{}/xy$	(5.20)

Note that the two Subsorts $\mathbf{F}_0$ and $\mathbf{F}_1$ are in fact identical, owing to the classifying property of $\hat{}$, as are the other two for derived domains; also, since the domain is assumed perfect, all the elements of all the Subsorts are identical except as between different formulae. That is to say, in an operational table for **F**, only two non-identical entries can appear if $\in \mathbf{R}_1\,\mathrm{D}$, or three if $\in \mathbf{R}_0\,\mathrm{D}$. It is obvious that the range of different functions that can be defined is very limited. There are, besides these, functions whose values (for non-identical arguments) are restricted to be blurs, of which most fail to pass the conditions imposed by Theorem (5.23) below.

In fact, all bivalent functions over a perfect Sort can be reduced to four classes, with characteristics specified as follows:

Gi: Subsort $\mathbf{F}_3$ singular ($=i\,\mathrm{G}iii$)
Hh: Subsort $\mathbf{F}_3$ restricted ($=\mathrm{H}\mathbf{h}xx, x(\in \mathbf{h}\,x)$)
J: Subsort $\mathbf{F}_2$ empty; $=\mathrm{J}xy\,\hat{\cup}\,xy$.
K: Subsort $\mathbf{F}_2$ empty; $=\mathrm{K}xy\,\mathrm{kD}$.

It remains to examine the properties of these functions.

5.H. VALUES OF THE FUNCTIONS

Since in every case the value of a formula $\mathrm{F}xy$ must depend on nothing but the arguments x, y, and since these arguments are twins, we have the following cases: (1) $\mathrm{F}xy$ does not depend on the separate values of x, y, but does depend on whether they are identical; (2) F*xx* depends on *x*, but F*xy* does not depend on *x*, *y*; (3) F*xy* does depend on the actual values of *x*, *y* in all cases. To express the particular values involved, we shall need first to define what I call the *complementary blur* of any element:

∈**S** D ⇒ = ᵕ̀*x* ,*u*(ǂ *x*,*u*(∈ D̄*u*)) (5.21) Df. ᵕ̀

so that ᵕ̀*x* is the smudge *k*`D/*x* of all elements of D *other than x*.

We first note that, because of the restricted concurrence of the symbols, the value of the expression ᵕ̂ G*ixx* G*iyz* cannot depend on any of the arguments, and must therefore be *k*D (which I shall hereafter regularly abbreviate to *k*). Since by definition = G*iii i* we are left with two possible functors of this class: = G*ihh* ᵕ̀ *i*, = G*ihi i* and = G*ijj i*, = G*iji* ᵕ̀ *i*; of these, the former does not satisfy Theorem (5.23.2), leaving the latter as our only functor of this type (1). Note that the functor G*i* does not effectively depend on *i*; any other symbol for an element of the domain could be substituted for *i* and all we get us a different spelling for the same function, for all the elements are twins. The symbol *i* in G*i* is thus not a true argument under G, and I shall not persist in writing it; but I shall continue to use the symbol *i* for the value of G*x*, *x*(∈ D*x*).

For any bivalent functor for which $\mathbf{F}_3$ has more than one member, F*xx* is in effect a univalent functor with argument *x*, that is to say, an endomorphism over D, and so we know from Theorem (5.16) that its values are restricted to *x* itself and *k*. As for F*xy*, its value can be *k* or, since we can write '' ˆ*xyz* as '' *pfp* where *f* is any functor giving an elementary value for the pair *p*, we can have *f p* as the value of F *xy*; the only possible *f* here is k, and = k*p* ᵕ̂ *xy*, so that we can also have = F*xy* ᵕ̂ *xy*. Combining these two cases we get therefore four possible functors, which, together with the G defined above, give the following table of results (the symbol **Ih** is included only in this table, since = **Ih***xy* JH**h***xx*H**h***yy* so that it is not an independent functor):

		G*xy*	H**h***xy*	**Ih***xy*	J*xy*	K*xy*
= *xy*	= *i x*	*i*	*k*	*k*	*x*	*x*
	∈**h** *x*	*i*	*x*	*x*	*x*	*x*
	'∈ᵕ **h**/*ix*	*i*	*k*	*k*	*x*	*x*
ǂ *xy*	∈**h**ᵕ̂ *xy*	ᵕ̀ *i*	*k*	ᵕ̂ *xy*	ᵕ̂ *x̂y*	*k*
	'∈**h**ᵕ̂ *xy*	ᵕ̀ *i*	*k*	*k*	ᵕ̂ *xy*	*k*

(5.22)

Although I shall not refer again to **Ih**, I shall continue to cite **Hh** as one of our bivalent functors, since it is not reducible to the others as **Ih** is; however, **Hh** will also be found redundant in due course.

5.J. PROPERTIES OF THE FUNCTIONS

Certain properties can be proved to hold for all bivalent functors defined over perfect Sorts, from which we may infer that any functor not exhibiting these properties does not exist over such domains.

First, directly from (4.2.2), we can prove that:

$$\in \mathbf{R}\,\mathrm{D},\quad \in \mathrm{D}\,\hat{\cup}\,\hat{\cup}\,xyz,\quad \in^{!}\bar{\mathrm{D}}\,\hat{\wedge}\bar{\mathrm{D}}\bar{\mathrm{D}}\,\mathbf{F} \Rightarrow\; = \mathrm{F}xy\,\mathrm{F}yx \qquad (5.23.1)\ \text{Th.}$$

$$\Rightarrow\; = \mathrm{FF}xyz\,\mathrm{F}x\mathrm{F}yz \qquad (5.23.2)\ \text{Th.}$$

The former is obvious; the latter is proved by $=\mathrm{FF}xyz\ \mathrm{FF}yzx$ from Theorem (5.23.1), and by the same again, $=\mathrm{FF}yzx\,\mathrm{F}x\mathrm{F}yz$, Q.E.D. The requirement of associativity is a powerful means of rejecting bogus functors.

We can also show that all functors distribute over H, J, K but not over G; that is to say

$$\in \mathbf{R}\,\mathrm{D}, \in \mathrm{D}\,\hat{\cup}\,\hat{\cup}\,xyz, \in \hat{_4}\,\mathrm{GHJK}\ \mathrm{E}, \in \hat{_3}\,\mathrm{HJK}\ \mathrm{F} \Rightarrow\; = \mathrm{E}x\mathrm{F}yz\,\mathrm{FE}xy\mathrm{E}xz \qquad (5.24)\ \text{Th.}$$

The proof is by enumeration of cases. I shall next prove that:

$$\in \mathbf{R}\,\mathrm{D},\quad \in^{!}\bar{\mathrm{D}}\,\hat{\wedge}\mathrm{DD}\,,\mathbf{F}(=i\,\mathrm{F}x,x(\in \mathrm{D}x)) \Rightarrow\; > {}^{c}\mathrm{D}\,3 \qquad (5.25)\ \text{Th.}$$

Take first the case where ${}^{c}\mathrm{D} = 1$, and let i be the only member of D; then the functor J, whose only instance is $=\mathrm{J}iii$, satisfies the condition. Next, if ${}^{c}\mathrm{D} = 2$, the condition is fulfilled by $=\mathrm{F}\ \mathrm{G}$. Now let ${}^{c}\mathrm{D} = 3$, so that $=\mathrm{D}\,\hat{_3}\,ghi$. The condition now requires that for all x, $=\mathrm{F}xx\ i$, and for all x, y ($\nmid xy$), $=\mathrm{F}xy\,\grave{\cup}\,i = \hat{\cup}\,gh$. But if so, $\mathrm{FF}ghi\ = \mathrm{F}\,\grave{\cup}\,ii$ which by Theorem (5.9) $= \hat{\cup}\,\mathrm{F}gi\mathrm{F}hi$ $= \hat{\cup}\,\grave{\cup}\,i\,\grave{\cup}\,i = \grave{\cup}\,i$, whereas $\mathrm{F}g\mathrm{F}hi = \mathrm{F}g\,\grave{\cup}\,i = \hat{\cup}\,\mathrm{F}gg\mathrm{F}gh = \hat{\cup}\,i\,\grave{\cup}\,i = k$. Therefore $\nmid \mathrm{FF}ghi\,\mathrm{F}g\mathrm{F}hi$, contrary to Theorem (5.23.2), so that F cannot be a functor over the perfect Sort D. If now D has more than three members, it must have a Subsort with three members, over which the above result would recur, thereby proving that such a functor does not exist in any larger domain, Q.E.D.

Thus, the functor G exists non-trivially only for the perfect Sort of two members. This suggests that we should tabulate the existence of the various functors, or their absence, over various perfect prime Sorts. In Table (5.26) below, endomorphisms E exist at least trivially for all the domains. For the rest, + indicates that the functor exists, – that it does not, and other entries indicate its reducibility to other functors. I include the identity I and the trivalent functor L discussed in 5.K:

Functor & Valency	ᶜD = 0	1	2	3	4	5 or more
G 2	–	= I	+	–	–	–
Hh 2	–	= I	= J	(see note below)		
I 1	–	+	∈ **En**	∈ **En**	∈ **En**	∈ **En**
J 2	–	= I	+	+	+	+
K 2	–	= I	= J	+	+	+
L 3	–	= I	= GG	–	+	–

(5.26)

The functor **Hh**, where **h** stands for any subsort of the domain, can be expressed in terms of K and the functor Ç to be defined in 6.C, as follows:

$$\in \mathbf{R}\mathrm{D}, \quad \leqq \mathrm{D}\mathbf{h}, \quad \Rightarrow = \mathbf{Hh}xy\ \mathrm{K}\mathrm{Ç}\mathbf{h}x\mathrm{K}xy \tag{5.27}$$

on account of which we do not actually have to concern ourselves about H provided that K exists, which is in all non-empty domains.

5.K. TWO-ARGUMENT FUNCTIONS OVER DERIVED PERFECT DOMAINS

We have seen, in (5.20), that for derived domains no distinction can be made between the Subsorts $\mathbf{F}_2$ and $\mathbf{F}_3$, so that no function can take more than two different values, one for identical and one for non-identical arguments. This leaves us, of those listed above, only G and K; J also, being as it were the identity functor on two arguments, can be defined for derived Sorts.

Let us then ask whether in fact these functors do exist for, say, a perfect derived Sort of two members. These members are, by definition, non-elementary; but since they are twins, their composition must be the same. They must agree in the number of their members, and these each to each in the numbers of *their* members, and so on indefinitely. In the simplest case a derived Sort can be written as ˆˆ*pq* ˆ*rs*, with members ˆ*pq*, ˆ*rs*; and since $\hat{=}$ ˆ*pq* ˆ*rs* ˆ being a classifying functor, we have also alternative pairs such as ˆ*pr*, ˆ*ps*. Reference to the definition of the cardinal of a (perfect) Sort (5.12) shows that it makes this number as large as possible; that is, all these apparent 'pairs', notwithstanding their seeming incompatibility, will be counted as contributing members of the Sort. We therefore have to conclude that

$$\in \mathbf{R}_1\,\mathrm{D} \Rightarrow \neq 2\,{}^{\mathrm{c}}\mathrm{D} \qquad \text{(5.28.1) Th.}$$

With the corollary, from Theorem (5.25), that there is no function of the type of G definable for such Sorts; formally stated:

$$\in \mathbf{R}_1\,\mathrm{D}, \quad \Rightarrow {}'\in\,!\,\bar{\mathrm{D}}\,\hat{\wedge}\mathrm{D}',\,\mathrm{D}(\in \mathrm{D}i),\mathbf{F}(=i\,\mathrm{F}x,x(\in \mathrm{D}x)) \qquad \text{(5.28.2) Th.}$$

Thus, the only bivalant functors existing for any derived domain are J, K, and consequently H.

There may be an alternative proof of Theorem (5.28.1) from the want of any construction, employing the functors defined in 4.K, by which a derived Sort of precisely two members could be obtained from prime arguments. The proof given could be criticized for assuming that (5.12) necessarily applies for all perfect Sorts, rather than for prime Sorts only; I have however been unable to find any well-founded objection to (5.12) as stated, for surprise at (5.28.1) hardly amounts to that.

5.L. FUNCTIONS OF MORE THAN TWO ARGUMENTS

There is a functor L over three arguments, definable over the perfect prime Sort of four members, thus:

$$\begin{aligned} \in \mathbf{R}_{0}\,\hat{{}_4}\,wxyz \Rightarrow \in\,!\,\mathrm{D}\,\hat{\wedge}\hat{\wedge}\mathrm{DDD}\,\mathbf{L}, \; &= \mathrm{L}xxx\,x \\ &= \mathrm{L}xxy\,y \\ &= \mathrm{L}xyz\,w = \grave{\cup}\,\hat{\cup}\,\hat{\cup}\,xyz \qquad \text{(5.29) Df. } \mathbf{L} \end{aligned}$$

A trivalent functor, just as a bivalent one, must be associative over a perfect Sort; the proof is an obvious extension of that given for Theorem (5.23.2). We can use this property to show that L is the only plurivalent functor definable over any perfect Sort. Let T be a functor of valency 3.

The value of Txxx, from the symmetry, can only be x or $\grave{\cup}\,x$;

if $= \mathrm{T}xxx\,\grave{\cup}\,x$,	then	$\mathrm{T}ab\mathrm{T}ccc = \mathrm{T}ab\,\grave{\cup}\,c$
	but	$\mathrm{T}a\mathrm{T}bccc = \mathrm{T}atc \neq \mathrm{T}ab\,\grave{\cup}\,c$, so $\mathrm{T}xxx = x$.
The value of	$\mathrm{T}xxy$	may be x, $\grave{\cup}\,x$, $\hat{\cup}\,xy$, $\grave{\cup}\,\hat{\cup}\,xy$, y, $\grave{\cup}\,y$, or k;
as $= \mathrm{T}xxx\,x$,	then	$\mathrm{T}ab\mathrm{T}ccc = \mathrm{T}abc$

and if $= \mathrm{T}xxy\,x$, then $TaTbccc = \mathrm{T}acc \nmid \mathrm{T}abc$
if $= \mathrm{T}xxy \grave{\smile} x$ then $TaTbccc = \mathrm{T}a \grave{\smile} cc \nmid \mathrm{T}abc$
if $= \mathrm{T}xxy \hat{\smile} xy$ then $TaTbccc = \mathrm{T}a \hat{\smile} bcc \nmid \mathrm{T}abc$
if $= \mathrm{T}xxy \grave{\smile}\hat{\smile} xy$ then $TaTbccc = \mathrm{T}a \grave{\smile}\hat{\smile} bcc \nmid \mathrm{T}abc$
if $= \mathrm{T}xxy\,k$ then $TaTbccc = \mathrm{T}akc \nmid \mathrm{T}abc$
if $= \mathrm{T}xxy \grave{\smile} y$ then $TaTbccc = \mathrm{T}a \grave{\smile} bc \nmid \mathrm{T}abc$, so $= \mathrm{T}xxy\,y$.

The value of $\mathrm{T}xyz$ may be $\hat{\smile}\hat{\smile} xyz$, $\grave{\smile}\hat{\smile}\hat{\smile} xyz$, or k;
now as $= \mathrm{T}xyx\,y$ then $TaTbcba = \mathrm{T}aca = c$
and if $= \mathrm{T}xyz \hat{\smile}\hat{\smile} xyz$, $\mathrm{TT}abcba = \mathrm{T} \hat{\smile}\hat{\smile} abcba = \hat{\smile}\hat{\smile} abc = \hat{\smile}\hat{\smile} abc \nmid c$
if $= \mathrm{T}xyz\,k$, then $\mathrm{TT}abcba = \mathrm{T}kba = k \nmid c$, so
$= \mathrm{T}xyz \grave{\smile}\hat{\smile}\hat{\smile} xyz = w$.

Thus we have confined T to the definition of L in 5.29, whose complete associativity, not demonstrated above, can be proved easily by further examples. We have therefore that

$$\in \mathbf{R}_0\,\mathrm{D}, \quad = \mathrm{D} \hat{}_4\, wxyz \Rightarrow = \dot{\bar{\mathrm{D}}}\, \hat{}\hat{}\mathrm{DDD}/\mathrm{L} \qquad (5.30)\ \mathrm{Th.}$$

As a corollary, there is no irreducibly trivalent functor over the perfect Sort of five members (or more); for, if these were a, b, c, d, e, $\grave{\smile}\hat{\smile}\hat{\smile} abc = \hat{\smile} de$ and $\mathrm{TT}abcba = \mathrm{T} \hat{\smile} deba = \hat{\smile}\hat{\smile} cde \nmid c$, so that associativity fails in all cases. Since every larger perfect Sort contains one of five members, there is always some case where a trivalent functor proves non-associative. Therefore

$$\in \mathbf{R}_0\,\mathrm{D}, \quad > 4\,{}^c\mathrm{D} \Rightarrow = \emptyset /, \mathrm{F}(\in \dot{\bar{\mathrm{D}}}\, \hat{}\hat{}\mathrm{DDD}\,\mathbf{F}) \qquad (5.31)\ \mathrm{Th.}$$

In the case of a quadrivalent functor Q, we find (by reduction to the case of $\mathrm{T}xxy$ above) that $= \mathrm{Q}xxxy\,y$; and from symmetry we know that $\mathrm{Q}xxyy$ can only be $\hat{\smile} xy$, $\grave{\smile}\hat{\smile} xy$, or k. Thus, $QabbQbaaa = a$, whereas $QabQbbaaa = Qa \hat{\smile} abaa = \hat{\smile} ab$ or worse, which shows that Q must be non-associative, whatever other values it may be assigned.

It now follows from Theorem (5.23.2) that Q as thus defined cannot be a functor over the perfect prime Sort of four members, and since this is a Subsort of any larger one, it is also not a functor over any such Sort. Moreover, since we have assumed nothing about Q except that it has four arguments, the existence of any functor of five arguments would imply that, by identifying two of the arguments, such a quadrivalent functor could be constructed, so that this too is excluded. We have therefore proved that

$$\in \mathbf{R}_0 \mathrm{D}, \quad > 3\,{}^{c}\mathrm{D} \Rightarrow = \emptyset \text{!`} \bar{\mathrm{D}} \,\hat{:}\, \mathrm{D}4 \qquad (5.32)\ \text{Th.}$$

This theorem and the one before show that we have now completed the enumeration of functions definable over perfect prime Sorts. All this however applies only to *finite* Sorts; we have yet to enquire what may be the case for infinite Sorts, if any.

5.M. INFINITE PERFECT SORTS

Let Y_n be the perfect prime Sort of n members, and let Y_n^2 be defined by $= \mathrm{Y}_n^2 \ \grave{}\, \hat{\wedge}\mathrm{Y}_n \mathrm{Y}_n \mathrm{Y}_n$, that is to say, Y_n^2 is to contain all the pairs in the pair-Sort $\hat{\wedge}\hat{\wedge}\mathrm{Y}_n$ except those identical-pairs which, if Y_n is a prime Sort, are already members thereof because $= \hat{}xxx$ if $\in \mathbf{E}x$. Further let $= \mathrm{Y}_n^{x+1} \ \grave{}\, \hat{\wedge}\mathrm{Y}_n^x \mathrm{Y}_n^x \mathrm{Y}_n^x$; then if $= {}^{c}\mathrm{Y}_n^x \mathrm{C}(n, x)$, we shall have $\mathrm{C}(n, x+1) = \mathrm{C}(n, x)\,(\mathrm{C}(n, x) - 1)/2$, and we shall get the following table of cardinals for small n, x:

$x =$	1	2	3	4
$n = 2$	2	1	0	0
3	3	3	3	3
4	4	6	15	105

(5.33)

Iteration of the above rule for finding Y_4^{x+1} thus generates an infinite sequence of perfect Sorts, of which none after the first are prime, but all are finite. Moreover, both the Sorts so constructed, and their individual members, are distinct from each other; the collections of such Sorts up to a given cutoff are therefore all *ordinable* Sorts, and have the properties of Sets, so that it is permissible to apply the Axiom of Choice to the sequence, and infer the existence of a Set having one member taken from each of the successive Sorts Y_4^x. We can therefore use transfinite induction to infer from the sequence of cardinals a limit cardinal, which is evidently in this case $2^{2^{\aleph_0}}$.

Can we construct by analogous means a Sort with a smaller infinite cardinal? In the above construction, the elements of Y_n^x have the form of pairs of those of the preceding Sort; we could form a series in which the members of successive Sorts were of the form $\hat{}a\,\hat{}b\,\hat{}\ldots\,\hat{}xy$ and examination shows that in this case $\mathrm{C}(4, x) = 12.3^{x-3}$ $(x \geqq 3)$, giving the lower limit $2^{\aleph_0}$. If however we go one further, and devise a sequence in which the members of successive Sorts are successively larger multiplets of elements of Y_4, the value of $C(4, x)$ stops short at 15 for $C(4, 4)$. I have been unable to find any way in which we can generate from a perfect prime Sort an infinite Sort of cardinal $\aleph_0$, except to postulate that Subsorts of this cardinality exist in larger Sorts.

We have however proved that

$$\in \mathbf{R}_1 \, \mathrm{S} \Rightarrow \ddagger \emptyset /, \mathrm{S}(> \aleph_0 {}^{c}\mathrm{S}) \qquad (5.34)\ \text{Th.}$$

This is an important result, for we shall find that at least one class of infinite Sorts plays an important role in the application of the theory made in Part II. On the other hand, there seems little hope of building an extensive theory of infinite Sorts on this foundation. To prove almost all of the theorems on infinite Sets, it is necessary to use the Axiom of Choice *within* the Sets in question, and this is by no means guaranteed for us by its availability *between* the Sorts used in constructing series such as in (5.33). It is not even, so far as I can see, possible to state such an axiom (in any of its various forms) in relation to a class of indistinguishables. Thus the use we can legitimately make of these constructions is very limited; little more, in fact, than what follows from their mere existence. This, however, is already a substantial step in the context of triparitous theory.

5.N. OPERATIONAL TABLES OF THE FUNCTORS

I give below operational tables for the functors G, H, J, K, L over perfect prime Sorts of 2, 3, and 4 members, with their blurs to complete the closures. The blur functor $\hat{\cup}$ is omitted to save space. The two functors Gi, Gj are twins; either (but not both) may be used. For H I give one specimen only of the Subsort pseudo-argument.

Gi	*i*	*j*	*k*
i	*i*	*j*	*k*
j	*j*	*i*	*k*
k	*k*	*k*	*i*

(5.35.1)

K	*i*	*j*	*k*
i	*i*	*k*	*k*
j	*k*	*j*	*k*
k	*k*	*k*	*k*

(5.35.2)

Gj	*i*	*j*	*k*
i	*j*	*i*	*k*
j	*i*	*j*	*k*
k	*k*	*k*	*j*

(5.35.3)

J	*h*	*i*	*j*	*hi*	*ij*	*jh*	*k*
h	*h*	*hi*	*jh*	*hi*	*k*	*jh*	*k*
i	*hi*	*h*	*ij*	*hi*	*ij*	*k*	*k*
j	*jh*	*ij*	*j*	*k*	*ij*	*jh*	*k*
hi	*hi*	*hi*	*k*	*hi*	*k*	*k*	*k*
ij	*k*	*ij*	*ij*	*k*	*ij*	*k*	*k*
jh	*jh*	*k*	*jh*	*k*	*k*	*jh*	*k*
k	*k*	*k*	*k*	*k*	*k*	*k*	*k*

(5.35.4)

$H^\wedge hi$	*h*	*i*	*j*	*hi*	*ij*	*jh*	*k*
h	*h*	*k*	*k*	*k*	*k*	*k*	*k*
i	*k*	*i*	*k*	*k*	*k*	*k*	*k*
j	*k*	*k*	*k*	*k*	*k*	*k*	*k*
hi	*k*	*k*	*k*	*hi*	*k*	*k*	*k*
ij	*k*	*k*	*k*	*k*	*k*	*k*	*k*
jh	*k*	*k*	*k*	*k*	*k*	*k*	*k*
k	*k*	*k*	*k*	*k*	*k*	*k*	*k*

(5.35.5)

L*g*	*g*	*h*	*i*	*j*	*gh*	*gi*	*gj*	*hi*	*hj*	*ij*	*ghi*	*ghj*	*gij*	*hij*	*k*
g	*g*	*h*	*i*	*j*	*gh*	*gi*	*gj*	*hi*	*hj*	*ij*	*ghi*	*ghj*	*gij*	*hij*	*k*
h	*h*	*g*	*j*	*i*	*gh*	*hj*	*hi*	*gj*	*gi*	*ij*	*ghj*	*ghi*	*hij*	*gij*	*k*
i	*i*	*j*	*g*	*h*	*ij*	*gi*	*hi*	*gj*	*hj*	*gh*	*gij*	*hij*	*ghi*	*ghj*	*k*
j	*j*	*i*	*h*	*g*	*ij*	*hj*	*gj*	*hi*	*gi*	*gh*	*hij*	*gij*	*ghj*	*ghi*	*k*
gh	*gh*	*gh*	*ij*	*ij*	*g*	*k*	*k*	*k*	*k*	*k*	*k*	*k*	*k*	*k*	*k*
gi	*gi*	*hj*	*gi*	*hj*	*k*	*g*	*k*	*k*	*k*	*k*	*k*	*k*	*k*	*k*	*k*
gj	*gj*	*hi*	*hi*	*gj*	*k*	*k*	*g*	*k*	*k*	*k*	*k*	*k*	*k*	*k*	*k*
hi	*hi*	*gj*	*gj*	*hi*	*k*	*k*	*k*	*g*	*k*	*k*	*k*	*k*	*k*	*k*	*k*
hj	*hj*	*gi*	*hj*	*gi*	*k*	*k*	*k*	*k*	*g*	*k*	*k*	*k*	*k*	*k*	*k*
ij	*ij*	*ij*	*gh*	*gh*	*k*	*k*	*k*	*k*	*k*	*g*	*k*	*k*	*k*	*k*	*k*
ghi	*ghi*	*ghj*	*gij*	*hij*	*k*	*k*	*k*	*k*	*k*	*k*	*g*	*k*	*k*	*k*	*k*
ghj	*ghj*	*ghi*	*hij*	*gij*	*k*	*k*	*k*	*k*	*k*	*k*	*k*	*g*	*k*	*k*	*k*
gij	*gij*	*hij*	*ghi*	*ghj*	*k*	*k*	*k*	*k*	*k*	*k*	*k*	*k*	*g*	*k*	*k*
hij	*hij*	*gij*	*ghj*	*ghi*	*k*	*k*	*k*	*k*	*k*	*k*	*k*	*k*	*k*	*g*	*k*
k	*k*	*k*	*k*	*k*	*k*	*k*	*k*	*k*	*k*	*k*	*k*	*k*	*k*	*k*	*g*

(5.35.6)

CHAPTER VI

REPRESENTATIONS OF INITIAL SORTS

In this chapter I shall begin to work out the ideas contained in Chapter III in a rigorous manner, using the material of Chapter V to do so. The aim is to construct, for as many Sorts as possible, a Biparitous Representation (hereafter often abbreviated as Bip. Rep.) satisfying the conditions stated in (3.2). As I have already explained, any Sort which lacks such a representation can have no epistemological status in the empirical world; and as most Sorts fall into this class, those which have representations must be expected to have some recognisable reflection in the world we observe. In searching for representations, I confine attention to prime Sorts, as defined in (5.19.1); derived Sorts (such as pair-Sorts, power-Sorts), being constructed from prime Sorts, are evidently representable if and only if their prime 'factors' are so. But among prime Sorts, there is a further distinction to be made.

6.A. INITIAL AND SUPERSTRUCT SORTS

A Sort which is defined without reference to any other Sort I shall call *initial*; one on the other hand which is defined by some construction or procedure involving other Sorts, already defined, will be call *superstruct*. Superstruct Sorts may be either derived or prime; I shall consider in detail only the latter, for the reasons stated. This will be the subject of Chapter VII; here I deal with initial Sorts only. It will sometimes be convenient to describe the Sorts S_1 S_2 ..., used directly in the definition of a superstruct Sort T as a function of S_1 S_2 ..., as the *substructs* of T.

Whenever a given function or set of functions is defined over a given Sort, I shall follow normal usage in calling the latter the *domain* of the functions. When a Sort is described as a domain without reference to any particular function, it will be understood to refer to all the functions which can be defined over the given Sort. I have shown in the last chapter that the range of functions definable over a given perfect Sort is quite restricted, and can in general be completely enumerated, as has been shown. Since an initial prime perfect Sort is defined merely by citing its cardinality, and since this cardinal by itself determines the functions for which the Sort is a domain, the existence of these functions is an essential and characteristic property of the Sort; this is why we can simply call it a 'domain', thereby implying that all these functions go along with it.

The situation is different with superstruct Sorts. In the first place, such Sorts are normally constructed from their substructs with the help of Bip. Reps., the competence of which may limit what functions can be defined for the superstructs. Again, a superstruct Sort is not defined merely by its cardinality, and its functions if any may be related somehow to those defined for its substructs; and new kinds of function can arise which have no basis in the substructs at all. It is possible to have a variety of different superstruct Sorts with the same cardinal. Most of these will be derived, and therefore of no immediate concern here; but others may be prime, and yet differ in their functions on account of how they are constructed. Such Sorts are in one sense identical; but they differ as to domains, and I adopt the view (which however is only marginally consequential) that where they both have valid Bip. Reps., they may be entitled to separate empirical interpretations.

6.B. COMPLEXES ON AN INITIAL SORT

Following rule R1.1.1 of (3.2), every element of a Bip. Rep has to be expressible as a formula involving only elements and functors of the Sort represented. Such formulae may differ either in their functors or in their arguments, or both. If the functors differ while the arguments remains the same, what they will represent if anything is the Sort whose members are the functors: in general, therefore, a given representation will be composed of formulae involving the same functor with different arguments. That being so, what particular functor is used is a matter of indifference, and the essence of the matter will be conveyed by the argument-strings alone.

I accordingly define a *complex of the first order* over a triparitous Sort S as any member of the class **Cx**S defined by

$$\in \mathbf{S}\mathrm{S},\ \in \mathbf{I}m \Rightarrow\ \leqq \mathbf{Cx}\,\mathrm{S}\,/\hat{m}, p_1(\in \mathrm{S}p_1)\ldots, p_m(\in \mathrm{S}p_m) \qquad (6.1)\ \mathrm{Df.}\,\mathbf{Cx}$$

while a complex of the *n*th order is one whose components p_i are themselves complexes of the $(n-1)$th order instead of members of S itself, as in (6.1). Note that we can only use the unordered multiplet-functor $\hat{m}$ in (6.1), since the p_i are members of S, which, in our applications (though not necessarily in general) will be a perfect Sort. For complexes of higher order however the components may be distinct and not twins, and in such cases ordered multiplets can be used (as will be done eventually in Chapter VII). A complex of the first order is simply itself a multiplet; the simplest non-trivial type is thus an unordered pair of members of the object Sort.

We have seen, as shown in Section 5.C, that in general perfect Sorts are not closed under the functions definable over them; but that their closures, defined by means of the blur functor $\hat{\cup}$, are so. Thus, it will be primarily the closure $\bar{\mathrm{D}}n_0$ of each perfect Sort of n members $\mathrm{D}n_0$ that we shall need to find representations for. And we have seen that ${}^c\bar{\mathrm{D}}n_0 = 2^n - 1$. Now it so happens, as will be proved in due course, that among all the n^m multiplets over $\mathrm{D}n_0$ we can find just $2^n - 1$ which form an ordinable Sort, each member being distinct from every other. This then is the maximum set available for forming a biparitous representation, and in each case all of them will be needed to represent the closure $\bar{\mathrm{D}}n_0$. It would seem therefore that in all cases a representation should be available.

However, we have not only to assign a distinct multiplet to each element of the closure, but we have also to be able to represent all the functions defined for the given domain (R2.3.1) and reproduce whatever symmetries may be exhibited by them (R2.3.2). These additional conditions restrict the variety of Sorts which have valid representations.

6.C. REPRESENTATION OF $\bar{\mathrm{D}}2_0$ BY PAIRS

The simplest non-trivial perfect Sort is $\mathrm{D}2_0$ the perfect Sort of two members. Its closure $\bar{\mathrm{D}}2_0$ contains $2^2 - 1 = 3$ members, the two sharp members (which I shall write i, j) and their blur, which is for this domain also the smudge, $\hat{\cup}\, ij = k$. In using letter symbols such as i, j for twin elements, it is important to remember the limitations of this usage. The symbol i stands for any sharp element of the domain, and a second symbol such as j is only meaningful in contrast with i. Any statement containing j and not i means exactly the same as if it had had i in place of j. Moreover, it was proved in Theorem (5.7) that $\hat{=}\, i\, k$, from which it follows that even though k is said to represent a blur, it only does so in the presence of i or j (or in the context of a sentence presupposing their existence) and in fact the whole domain $\bar{\mathrm{D}}2_0$ could be equally well represented with i, k sharp and j as their blur. This matter will be gone into in detail in 6.F.

There are six unordered pairs which can be derived from $\bar{\mathrm{D}}2_0$, namely $\hat{}ii$, $\hat{}ij$, $\hat{}jj$, $\hat{}ik$, $\hat{}jk$, $\hat{}kk$, of which $\hat{=}\, \hat{}ii\, \hat{}jj$, $\hat{=}\, \hat{}jj\, \hat{}kk$, and $\hat{=}\, \hat{}ik\, \hat{}jk$. A set of three distinct pairs is $\hat{}ii$ (usually written $/i$), $\hat{}ij$, and $\hat{}ik$. $\dot{-}/i\, \hat{}ij$ because they differ in cardinality, while $= \hat{}ik\, \hat{\cup}/i\, \hat{}ij$ which as a blur of distinct elements is itself distinct from either, so that $\dot{-}\, \hat{}ij\, \hat{}ik$.

More generally, it will be noted that two complexes of the same multiplicity will be distinct if they have different compositions in terms of blurs: that is to say $\neq \hat{4}\, i\, j\, \hat{\smile}\, jh\, k \;\; \hat{4}\, i\, \hat{\smile}\, ij\, \hat{\smile}\, jh\, k$, since the first has two sharp elements, one blur of two, and one smudge; whereas the second has one sharp, two blurs of two, and a smudge. It is easy to see that, in a perfect domain with three or more sharp elements, $= \hat{3}\, i\, j\, \hat{\smile}\, jh \;\; \hat{3}\, ijk$, since the symbol '*h*' denotes any other element besides *j*; it is therefore helpful to represent all elements of a multiplet in their sharpest form. Any multiplet can then be represented by a string of numerals, beginning with 1 and proceeding by steps of 0 or 1 but not more, and ending before the number of digits exceeds the multiplicity assigned to the multiplets. Any two nonidentical multiplets having the same formula in this notation will be twins, and any two having different formulae will be distinct. In the case of $\bar{\mathrm{D}}2_0$ we can set out the representation of the elements as follows:

$$
\begin{array}{llll|lll}
= \mathbf{Rp}\, i & /i & \text{or} & 1 & = \mathbf{Im} & 1 & i \\
= \mathbf{Rp}\, j & \hat{}ij & \text{or} & 11 & = \mathbf{Im} & 11 & j \\
= \mathbf{Rp}\, k & \hat{}ik & \text{or} & 12 & = \mathbf{Im} & 12 & k
\end{array}
\qquad (6.2)
$$

The formulae on the right, using the 'functor' **Im**, are strictly nonsense, since they purport to define the mapping of an ordinable Sort namely $\bar{\mathrm{H}}2_0 = \hat{3}\ 1\ 11\ 12$ onto a perfect Sort, $\bar{\mathrm{D}}2_0$, which according to Theorem 11 (5.4.3) is a smudged mapping. Nevertheless, the improper functor **Im** is useful at times as a notional inverse of **Rp**, provided that no consequences are inferred from statements containing it.

6.D. REPRESENTATION OF FUNCTORS OVER $\bar{\mathrm{D}}2_0$

To satisfy the requirements of (3.2) for constructing a valid representation of $\bar{\mathrm{D}}2_0$ we have to be able to represent, still using formulae using only symbols defined within $\bar{\mathrm{D}}2_0$, all the functors definable over the domain. There are in effect only two of these, namely G and J (there is also a third functor twin to G, but it denotes the same function and so can be disregarded). For the elements the representation given in (6.2) is unique; I shall use the symbols $\hat{\mathrm{G}}$ and $\hat{\mathrm{J}}$ for the representations of G, J which we shall try to find.

The first step is to arrive at a definition for the meaning of F $\hat{}wx\ \hat{}yz$ where F is defined over $\bar{\mathrm{D}}2_0$ and *w*, *x*, *y*, *z* denote elements of the same domain. First, we note that although $\hat{}$ denotes an unordered pair, if we adhere to the rule that *j* may not appear except in company with *i*, and that *k* always stands last, formal ambiguity of ordering is avoided; moreover, since in the formula *y*, *z* are non-concurrent with *w*, *x*, this rule applies separately for each argument. This being so, the statement $= \mathrm{F}\ \hat{}wx\ \hat{}yz\ \hat{}\mathrm{F}wy\mathrm{F}xz$ is unambiguous, and enables us to propose particular definitions, of which the simplest are:

$$\in \bar{\mathrm{D}}2_0\ \hat{\cup}\hat{\cup}\hat{\cup}\, wxyz, \quad \in !\bar{\mathrm{D}}2_0\ \hat{}\bar{\mathrm{D}}2_0\ \bar{\mathrm{D}}2_0\ \hat{\cup}\,\mathrm{GJ}, \ \in !\bar{\mathrm{H}}2_0\ \hat{}\bar{\mathrm{H}}2_0\ \bar{\mathrm{H}}2_0\ \hat{\cup}\,\hat{\mathrm{G}}\hat{\mathrm{J}} \Rightarrow$$
$$= \hat{\mathrm{G}}\ \hat{}wx\ \hat{}yz\ \hat{}\mathrm{G}wy\mathrm{G}xz,$$
$$= \hat{\mathrm{J}}\ \hat{}wx\ \hat{}yz\ \hat{}\mathrm{J}wy\mathrm{J}xz \qquad (6.3)$$

For example $\hat{\mathrm{G}}/i\ \hat{}ij = \hat{}\mathrm{G}ii\mathrm{G}ij = \hat{}ij$; $\hat{\mathrm{G}}\ \hat{}ik\ \hat{}ik = \hat{}\mathrm{G}ii\mathrm{G}kk = \hat{}ii$. It is evident, in this simple case, that the Ĝ, Ĵ thus defined do in fact give the required products for all the three elements of $\bar{\mathrm{D}}2_0$. This is shown in the following combined table, where Ĝ is given above the diagonal, and Ĵ below it:

Ĝ Ĵ	1 (*i*)	11 (*j*)	12 (*k*)
1 (*i*)	1 (*i*) 1 (*i*)	11 (*j*)	12 (*k*)
11 (*j*)	12 (*k*)	1 (*i*) 11 (*j*)	12 (*k*)
12 (*k*)	12 (*k*)	12 (*k*)	1 (*i*) 12 (*k*)

(6.4)

6.E. REPRESENTING THE SYMMETRY OF THE FUNCTIONS

We also have to check that the representation we have found satisfies the rule R3.2.2 concerning symmetry. Ĝ, like G, defines a group structure on $\bar{\mathrm{H}}2_0$; but whereas no problems arise from the need to choose a particular element of $\bar{\mathrm{H}}2_0$ as the unit of the group, in the case of $\bar{\mathrm{D}}2_0$ which is a perfect Sort all three choices should be equivalent. In fact, it is obvious that if *i* is taken as the unit, the elements *j*, *k* behave similarly to each other, and therefore that with *j* as the unit, *i*, *k* would be similar too. It ought to follow that *k* could be chosen as the unit; and indeed it could, but not without a corresponding adjustment in the table for Ĵ.

For unlike Ĝ, Ĵ defines a lattice structure (missing one of the bounding elements) on $\bar{\mathrm{H}}2_0$ reflecting that of $\bar{\mathrm{D}}2_0$ under J; in this structure, the blur k is distinguished from the sharp elements i, j. But since $\bar{D}2_0$ is a perfect Sort, any pair of its elements can be chosen as being 'sharp', with the third being the blur, and this last is not *simultaneously* eligible as the unit of the G-group. Thus, if we select the elements represented by 1, 12 as being 'sharp', the remaining symbol 11 must represent their blur, and now the unit of the G-group must be either 1 or 12. The case of $\bar{D}2_0$ is partly trivial in this respect, since it has not enough elements to fail the symmetry test of R2.3.2; this will require further study when we come to $\bar{D}3_0$.

Nevertheless, the representation (6.2) has been shown to satisfy the requirements for a valid biparitous representation $\bar{H}2_0$ of $\bar{D}2_0$. We may note in passing that $D1_0$ (which is its own closure) is trivially represented by setting Rp i = /i; since this Sort has only one member, it is both perfect and ordinable. Its only functor is the identity I, and its representative $\hat{I}$ is the identity on $H1_0$ = //i. There are no symmetry requirements. Finally, D0 with no elements is obviously represented by the empty Set $\emptyset$. Thus we have valid Bip. Reps for D0, $D1_0$, and $\bar{D}2_0$.

6.F. REPRESENTATION OF $D3_0$

It is now clear that to represent this Sort we shall require to use triplets instead of pairs. There are seven elements of $\bar{D}3_0$ which we may denote in alphabetic symbols as $\hat{7}$ *i j h* ˆ*ij* ˆ*jh* ˆ*hi k*. From these we can form 84 non-identical triplets, of which the largest ordinable Subsort contains 7, which we can denote, using the numerical notation described above, by 1 11 111 12 112 122 123. Each of these can be represented by a triplet of letter symbols, with the exception of 122, respectively /*i* ˆ*ij* $\hat{3}$ *ijh* ˆ*ik* $\hat{3}$ *ijk*( ) ˆ*ikk*; the *best we can do for* 122 *is* ˆı $\hat{\circ}$/*i* ˆ*jh* which is the blur of /*i* for 1 and $\hat{3}$ *ijh* for 111.

There are always precisely the right number of the numeral symbols to provide distinct multiplets to represent the elements of any perfect Sort. There is one with one digit, 1; two with two digits 11, 12; four with three digits 111, 112, 122, 123; then 1111, 1112, 1122, 1222, 1123, 1223, 1233, 1234, and so on. In effect this representation of the elements is unique, so that we have no choice when we find one that fails for the functors or for the symmetry conditions.

In the case of $\bar{D}3_0$ we have therefore the representation $\bar{H}3_0$:

= **Rp** *i* 1	= **Rp** ˆ*ij* 12		
= **Rp** *j* 11	= **Rp** ˆ*ih* 122	= **Rp** *k* 123	(6.5)
= **Rp** *h* 111	= **Rp** ˆ*jh* 112		

or some permutation of this – which raises more problems than did the permutation of representatives for $\bar{D}2_0$.

6.G. REPRESENTATION OF FUNCTORS OVER $\bar{D}3_0$

As we saw in Chapter V, there are three bivalent functors over this domain, namely H, J, K; and also a Sort of univalent Endomorphisms which I shall look into in Chapter VII. The functor H can be reduced to J and K by (5.27), so that it needs no separate consideration.

To find a representation for J, we can use the same formula (6.3) which we used for $\bar{D}2_0$, modified to allow for the use of triplets. This gives us, provided the alphabetic precedence rule is used,

$$\in \bar{D}3_0 \,, p(\in \hat{6}\, uvwxyz\, p) \Rightarrow = \hat{J}\, \hat{3}\, uvw\, \hat{3}\, xyz\, \hat{3}\, \mathrm{J}ux\mathrm{J}vy\mathrm{J}wz \qquad (6.6)$$

For example, $= \mathbf{Rp}(\mathrm{J}h\, \hat{\smile}\, jh)\hat{\mathrm{J}}\, \hat{3}\, ijh\, \hat{3}\, ijk$ (otherwise $\hat{\mathrm{J}}$ 111 112), which by (6.6) becomes $\hat{3}\, \mathrm{J}ii\, \mathrm{J}jj\, \mathrm{J}hk = \hat{3}\, ijk = \mathbf{Rp}\, \hat{\smile}\, jh$. The calculations can be greatly helped by the following rule, which enables any numerical symbol to be interpreted as a blur of strings of 1's, which are understood to denote sharp elements, as in (6.5). The rule is that a new digit appears after the completion of any string of 1's included in the blur. Thus, for quadruplets, $1122 = \hat{\smile}\, 11\ 1111$ because the new digit 2 appears after two, giving 11, and new (null) digit appears again after four, giving 1111; and $\hat{\smile}\, 111\ 1122 = 1123$, and so forth. The rationale of this rule is contained in (6.6) (remember that for *initial* Sorts J is synonymous with $\hat{\smile}$).

The operational table for $\hat{\mathrm{J}}$ which emerges from (6.6) for $\bar{D}3_0$ is in part as follows:

J	1	11	111	12	112	122	123
1	1	12	122	12	123	122	123
11	12	11	112	12	112	123	123
111	122	112	111	123	112	122	123

(6.7)

The reader may check that the entries in this table accord with the rule for decomposition given above. It will be noted that it is isomorphic with the first three rows of the table for J given in 5.N, showing that the representation is satisfactory.

The functor K can be represented by $\hat{\mathrm{K}}$ defined identically with $\hat{\mathrm{J}}$ as in (6.6), but with the alphabetic precedence rule *not* applied. This gives for identical arguments $\hat{\mathrm{K}}pp = \hat{\mathrm{J}}pp = p$, but in the general case

$$\in \bar{\mathrm{D}}3_0 \,, p(\in \hat{6}\, uvwxyz\, p) \Rightarrow\ = \hat{\mathrm{K}}\ \hat{3}\, uvw\ \hat{3}\, xyz\ \hat{3}\, \mathrm{J}ux\mathrm{J}vy\mathrm{J}wz \tag{6.8}$$

the non-concurrence between u, v, w and x, y, z allows random substitution of values, so that the triplet of J's can be anything, i.e. $= k$. Thus all the bivalent functors defined over $\bar{\mathrm{D}}3_0$ have satisfactory representations, as well as all the elements. We have now to examine the symmetry conditions required.

6.H. SYMMETRY OF THE FUNCTIONS OVER $\bar{\mathrm{D}}3_0$

The structure imposed on $\bar{\mathrm{D}}3_0$ by K is trivial, and in any case subsumed in that deriving from J. The latter gives to the domain the structure of a lattice, missing one of its bounds; and since $\bar{\mathrm{D}}3_0$ is a perfect Sort, all mappings of its seven elements onto the points of the lattice are equally valid. If this fact is to be conveyed in terms of a representation of the Sort by a biparitous Set, it must be possible to choose any three elements of the representation as representing 'sharp' elements, and then to interpret unambiguously any other representative element as the blur of a given collection of sharp elements. If there is ever any ambiguity, the representation fails to satisfy the rule R1.1.2 with the given choice of 'sharp' elements; so that, if it does satisfy R1.1.2 with any other choice, it is in breach of the required symmetry. As we have seen in (6.5) that this rule *is* satisfied when the sharp elements are represented by 1 11 111, we have to check that it holds equally in other cases.

This check is best applied by the use of letter symbols. I set out, below the elements of $\bar{\mathrm{D}}3_0$ in a row of three sharp elements, a lattice diagram without its upper bound. The top row can be chosen in seven ways (one of which has h, i, j for the sharp elements, as in (6.5)); in every case there is a simple rule for finding the other two rows, and therefore for assigning the representatives in numerical code by reading off from (6.5).

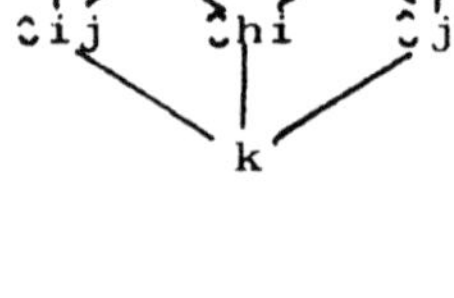

(0)

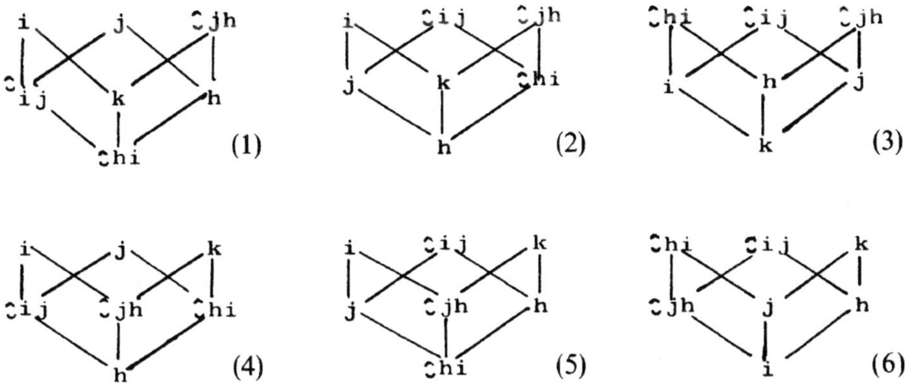

The rule for forming blurs, from the symbols in the first row, in all cases except No. (3) is to retain all and only those letters which occur an odd number of times in the top row: thus, in (1), from *j* & ˆ*jh* we have 2 *j*'s and 1 *h*, so their blur is given as *h*; and from *i*, *j*, ˆ*jh* we have 1 *i*, 2 *j*'s, and 1 *h*, giving ˆ*hi* in the third row. In (3) the rule is just the opposite, we retain those which occur an even number of times; this case is marked by the appearance of the original first row as the second, which happens nowhere else.

There are thus no ambiguities in any permutation of the representation for the mapping from $\bar{D}3_0$ onto $\bar{H}3_0$; and of course total ambiguity in every permutation for the reverse mapping; both situations being in accord with the required symmetry among all permutations. The condition R2.3.2 is thus met for $\bar{D}3_0$, as well as R2.1, R2.2, and R2.3.1, thereby proving that this domain has a valid representation, along with D0, $D1_0$, and $\bar{D}2_0$.

6.J. REPRESENTATION OF $\bar{D}4_0$

The next initial perfect Sort is $D4_0$ with four elements, with its closure $\bar{D}4_0$ having 15 elements including blurs. It is evident from the examples afforded by the preceding domains that these 15 elements can be represented by numerical codes using up to 4 digits, the four sharp elements being 1 11 111 1111. $\bar{D}4_0$ is closed under three bivalent functors H, J, K and also under the trivalent L which exists only for this domain. Representations $\hat{J}$ and $\hat{K}$ analogous to those used in $\bar{D}3_0$ can obviously be constructed for J, K; and H can be defined in terms of these. It remains to consider L.

As we see from (5.2) defining L, we have = L*wxyz*. Thus, $\hat{\text{L}}$ 1 11 111 = 1111; and also, $\hat{\text{L}}$ 11 111 1111 = 1. In letter symbols this becomes $\hat{\text{L}}$ $\hat{\ }ij$ $\hat{3}\,ijh$ $\hat{4}\,ijhg$ = $/i$. It is clear that no one rule of derivation will work for both formulae. It is also the case that L cannot be reduced to formulae using bivalent functors, so there is no alternative but to construct $\hat{\text{L}}$ in terms of L. This being impossible, the representation fails.

This should be enough to show that $\bar{\text{D}}4_0$ has no valid biparitous representation. But it also fails in respect of the symmetry conditions on the representation of J. This is easily shown: for if we choose $\hat{\smile}\,gh$, $\hat{\smile}\,hi\,\hat{\smile}\,ij\,\hat{\smile}\,jg$ to represent the four sharp elements, there is no formula which can assign different symbols to the blurs of $\hat{\smile}\,gh$, $\hat{\smile}\,ij$ and of $\hat{\smile}\,hi\,\hat{\smile}\,jg$; but these are nevertheless non-identical elements of $\bar{\text{D}}4_0$, though with this choice of sharp elements no representation can avoid ambiguity at this point (and some others), though there is none if the sharp elements are represented by $/i$, $\hat{\ }ij$, $\hat{3}\,ihj$, and $\hat{4}\,i\,jhg$. We have thus proved that

$$\in \mathbf{R}_0\,\text{D}n_0, \Rightarrow 4\,{}^c\text{D}n_0, \nmid \emptyset/, \text{H}(= \mathbf{Rp}\,\bar{\text{D}}n_0\,\text{H}) \Rightarrow (/) \qquad (6.9)$$

From this we have the corollary that, if any larger finite perfect Sort had a representation, a subset of this would serve to represent $\bar{\text{D}}4_0$, that is to say

$$\in \mathbf{R}_0\,\text{D}n_0, > 3\,{}^c\text{D}n_0, < \aleph_0\,{}^c\text{D}n_0, \nmid \emptyset/, \text{H}(= \mathbf{Rp}\,\bar{\text{D}}n\,\text{H}) \Rightarrow (/) \qquad (6.10)\ \text{Th.}$$

The corollary does not hold for the infinite case, however, for it is not possible to apply the symmetry condition meaningfully to such Sorts for reasons which are explained below.

6.K. REPRESENTATION OF INFINITE SORTS

It was shown in Chapter V that it is possible to construct at least one kind of infinite perfect Sort, which, in conformity with the notation of this chapter, I shall now call $\text{D}\infty_0$. This is not, like those we have been considering up to now, a prime Sort but a derived one, its members being almost all infinite complexes on the perfect Sort Y_4. Its functors are therefore at most H, J, K (Theorem 5.28.2). Now, both H and K have values, for nonidentical arguments, which are the smudge of their domain; the smudge of an infinite perfect Sort is however a function of an infinite number of arguments, and such a function cannot be defined unless the arguments can be ordered, which in this case they cannot be; therefore there is no such smudge, and the functors which produce it do not exist for this domain. (The nonexistence of $kD\infty_0$ is analogous to the nonexistence of an 'integer chosen at random' from all the integers.) We have therefore only the functor J to consider.

To represent the elements of $D\infty_0$ we shall obviously need to use infinite multiplets. There is no objection to this in principle; we hold that π can be represented by an infinite string of decimal digits, which admittedly form an ordered Set whereas up to now we have been using unordered multiplets. But there exist more general classes of complexes, including multiplets of distinct multiplets, which can also be ordered even if they are infinite strings. By this means we could, in principle, represent the members of $\bar{D}\infty_0$ including all the finite blurs. In our numerical code, the sharp members of $D\infty_0$ are represented by 1 11 111 &c. with no limit on the length of the string, and their blurs by the corresponding codes in which higher digits appear.

However, while in the case of finite Sorts there are only a finite number of permutations of sharp elements in respect of the 1-string codes, for each of which every other element of the closure has a unique representative, the permutations of the sharp elements of $\bar{D}\infty_0$ are not even denumerable, and consequently the rules, which we have seen can be provided for all cases with $\bar{D}3_0$ and for some only for $\bar{D}4_0$, cannot be formulated at all for $\bar{D}\infty_0$. Both the first and the last of these preserve the symmetry proper to a perfect Sort, as required by R2.3.2: $\bar{D}4_0$ and all subsequent finite Sorts fail to do so. Thus, for $\bar{D}\infty_0$ we have a biparitous representation which satisfies all requirements.

We have still to consider what exactly it is that we have represented. Three attitudes to this question might be defended. First, we might adopt a strict Intuitionist standpoint, and refuse to consider any *completed* infinity. This requires that the cardinality of our representation $\bar{H}\infty_0$ must be $\aleph_0$, and that we can therefore represent only a denumerable Subsort of $\bar{D}\infty_0$. Moreover, to anticipate Chapter VII, the Intuitionist position precludes the existence of any endomorphisms on an infinite perfect Sort, so that there is only one such ($D\infty_0$) which exists, and this is only partly representable. I shall call this Treatment A.

If we do not deny the legitimacy of considering completed infinities, we may still insist that any *representation* has to be in some sense realizable, and must therefore consist of finite complexes, even though there is no need to impose an upper limit on their length. This Treatment B still limits us therefore to representations of denumerable cardinality, but admits the existence of Sorts of endomorphisms on infinite perfect Sorts, of each of which a denumerable Subsort is representable. In this case we are allowed an infinite sequence of biparitous representations all of cardinality $\aleph_0$. However, the arguments leading to (3.2) where the requirements for a valid representation are codified apply equally well to imaginary representations, of which we need not insist that their terms must be writable. If this scruple is waived, we are still limited by the nonexistence of infinite blurs, the main consequence of which is that the closure of any infinite $D\infty_n$ has the same cardinality as $D\infty_n$ itself (I therefore denote it by $\breve{D}\infty_0$); but we are left free to consider a sequence of infinite domains, each having a valid representation and yet each having a greater cardinality than its precessor. This I shall call Treatment C.

The relevance of these three 'treatments' to the subsequent application of the theory will be discussed in Chapter VIII. It is possible to take up rigid philosophical positions which may lead to only one of them being acceptable; if my interpretation is correct, such a stance is equivalent to an a priori judgment on the nature of the world, which can hardly be a legitimate inference from the kind of presuppositions involved. None of these treatments leads to an actual mathematical contradiction (provided that Treatment C refrains from asking about the *limit* of the sequence of infinite Sorts), and so, I maintain, all can be 'entertained'—whatever that means.

6.L. CONSPECTUS OF REPRESENTATIONS OF INITIAL SORTS

I conclude this chapter with a summary of the results arrived at. We have examined all the initial perfect prime Sorts, and the initial perfect infinite Sort which is not prime, each uniquely characterized by its cardinal, and have found that:

D0 with no members, is represented by $\emptyset$ (6.11.1)

$D1_0$ with one member and the functor I, is represented by any singular Sort and the identity over that (6.11.2)

$\bar{D}2_0$ with two sharp members and one blur, and the functors G, J, is represented by the ordinable Subsort $\bar{H}2_0$ of $\hat{:}\bar{D}2_0 2$ and the functors $\hat{G}$, $\hat{J}$ as defined in (6.3) and (6.4) (6.11.3)

$\bar{D}3_0$ with three sharp members and four blurs, and the functors H, J, K, is represented by the ordinable Subsort $\bar{H}3_0$ of $\hat{:}\bar{D}3_0 3$, by the functors $\hat{J}$, $\hat{K}$ of (6.3), and $\hat{H}$ derived from these by (5.23) (6.11.4)

$\bar{D}4_0$ with four sharp members and 11 blurs, the bivalent functors H, J, K, and the trivalent functor L, has no valid representation satisfying (3.2) (6.11.5)

$\bar{D}n_0(>4n)$ with n sharp members and $2^n - n - 1$ blurs and the functors H, J, K, have no representations (6.11.6)

$\breve{D}\infty_0$ with $2^{\aleph_0}$ non-elementary sharp members and the same number of finite blurs, and the sole functor J, is represented
on Treatment A or B, as to a denumerable Subsort, by the ordinable Subsort $H\infty_0$ of $\hat{\because}D\infty_0\aleph_0, n(<\aleph_0 n)$, but
on Treatment C, completely by the ordinable Subsort $\breve{H}\infty_0$ of $\hat{\because}D\infty_0\aleph_0\aleph_0$,
and the functor J as for finite Sorts (6.11.7)

We have in this chapter paid no attention to the endomorphisms, univalent functors definable over perfect Sorts, which will be considered in Chapter VII. It will be found that these can also be represented for all those initial Sorts which have representations otherwise.

There remains one initial Sort not listed above, which may have to be considered. This I denote by D! and define as having every other representable Sort as a Subsort. For Treatment C D! does not exist; for the other Treatments its existence is no problem, but there is considerable doubt whether it can have a representation, or come to that whether its having or not having one makes any difference to its interpretation. The need for it is explained in 7.M.

CHAPTER VII

REPRESENTATION OF SUPERSTRUCT SORTS

In the preceding chapter I have examined the initial prime perfect Sorts, to determine which of them would have valid Bip. Reps., and I found this to be the case for those of cardinalities 0, 1, 2, 3 and ∞, but for no others, so that these might be rationally expected, by the arguments given in Chapter III, to have manifestations on the physical Plane. Contrasted with these initial Sorts, there are superstructs, defined as classes of operations on previously defined Sorts. In general, this term includes functions of either one or two arguments; but since the latter type involves derived Sorts, which I have already given reasons for not pursuing, I here concentrate on those superstructs composed of univalent functors, that is of endomorphisms. Sorts of endomorphisms were shown in Chapter V to be perfect prime Sorts, but are of course not initial Sorts; they will be the only type of superstructs to engage our detailed attention.

This chapter will first discuss briefly the case presented by Sorts of bivalent functors, but thereafter will survey the entire field of endomorphisms. When we have determined which Sorts of endomorphisms can have valid representations, we shall be able to draw up a complete list of 'rational' Sorts, in the sense of Chapter III, including both finite and infinite examples.

7.A. SORTS OF BIVALENT FUNCTORS

Any Sort which is derived by some well-defined function from two arguments, each of which is itself rational, must in turn have a Bip. Rep. formed from those of its arguments by (a representative of) the function concerned. The interpretation of such a rational Sort must obviously pay attention to the facts of its ancestry, and I shall assume that this is done by accepting only those interpretations which directly derive from those accepted for the arguments. For this reason, such derived Sorts will not contribute anything new to the total corpus of interpretations listed below in 7.N.

Here however we are concerned not with the products of such functions, but with the functors themselves. There is again little of interest in considering their Sorts, which when not ordinable are typically very small. The

Sort of bivalent functors over $D2_0$ is for example $\hat{3}GiGjJ$ having disjoint Subsorts $\hat{}GiGj$ isomorphic with $D2_0$ itself and /J. The bivalents over $D3_0$ include J and K which are distinct, and the class of functors **Hh** which is isomorphic with the endomorphisms over $D3_0$. It is arguable that these isomorphs are entitled to independent interpretations, just as $D2_1$ is from its isomorph $D3_0$; but I shall not pursue this here.

It is also possible to define multivalent functors by combination of bivalent ones. The values defined by these functors are complexes of various kinds; two functors are identical if and only if their values for identical arguments are identical, which in the present case means if the class of complexes generated are identical. Since complexes which differ in formal structure are distinct (except in degenerate cases), this means that the class of complex-generating functors is necessarily biparitous, and consequently does not concern us here.

One possible further complication needs attention here; namely, the possibility of constructing classes of endomorphisms over derived Sorts. However, of the two methods of constructing representations, which I shall discuss below, the only one generally available to represent such a Sort of endomorphisms is to use 'cedilla functors'; and it will be shown in 7.D that this expedient also fails. One can never be certain that a way round this difficulty might not be found: but meanwhile Sorts of endomorphisms on derived Sorts appear to have no Bip. Reps., and so do not qualify as rational. Our attention can therefore be confined to endomorphisms on initial prime and infinite derived Sorts, and to these I now turn.

7.B. FAMILIES OF ENDOMORPHISMS

The endomorphisms on any perfect Sort (considered as functors, not as mappings) constitute among themselves a perfect Sort (5.F). This Sort can only have a Bip. Rep. if its immediate substruct has one, for without this we have no means of referring to the members of the Sort on which the endomorphisms operate. Though necessary, this condition is not sufficient, for we still have to find the conditions under which an appropriate Bip. Rep. can be actually constructed.

If, however, we find such a representation, it may be possible to find one also for the endomorphisms *on* the Sort of endomorphisms, that is, for a second-level superstruct; and so on, it may be, indefinitely. This iterative procedure results in a sequence of rational Sorts, each containing endomorphisms on its precessor, which I shall call a *family* of endomorphisms.

Before giving a formal definition of this term, however, I have to show that the Sort of endomorphisms on any perfect substruct Sort is not only perfect but also prime. For we have seen that derived Sorts (except in the infinite case) will not in general give rise to endomorphisms having Bip. Reps.

I shall now show that

$$\in \mathbf{R}_0\,\mathrm{D} \Rightarrow \in \mathbf{R}_0\,\mathbf{En}\,\mathrm{D} \qquad (7.1)\ \text{Th.}$$

The individual members of **En** D are not mappings, but functors; that is, they are not defined as classes of ordered pairs, but as operators which either do or do not change their arguments into the smudge kD. There are therefore no Sorts, other than themselves, which can be presented as members of them; that is, $\leqq \mathbf{E}\,\mathbf{En}\,\mathrm{D}$, and from (5.19.1) the apodosis of (7.1) follows, Q.E.D.

Thus the only condition we have to place on the existence of **En** D is that D should have a Bip. Rep. If then **En** D also has a Bip. Rep., we have a further Sort **EnEn** D, and so on. It is evidently desirable to adopt a notation for such Sorts which will make it easy to see what place each one occupies in such a family. This I shall do by attaching subscripts to the symbols for the Sorts, indicating their *rank*:

$$\in \mathbf{R}\,\mathrm{D}n_r \Rightarrow = \mathrm{D}n_{r+1}\,\mathbf{En}\,\mathrm{D}n_r \qquad (7.2)$$

The value $r = 0$ will be reserved for initial Sorts. One might expect each initial Sort would generate a family disjunct from every other, but this turns out not to be the case. We shall find, for example, that $\mathrm{D}2_1$ and $\mathrm{D}3_0$ are actually identical, though one is defined as superstruct and the other as initial, which is claimed to justify the assignment of different interpretations to them. This imposes a need for care in defining a 'family'; I propose the following:

$$\in \mathbf{S}\,\mathrm{Z},\quad \in \mathbf{R}_0\,\mathrm{D},\quad \hat{\in}\,\mathrm{Z}\mathbf{En}{,}\mathrm{X}(\in(\cup\,\mathrm{Z}/\mathrm{D})\mathrm{X}) \Rightarrow = \mathbf{Fa}\,\mathrm{DZ} \qquad (7.3)\ \text{Df.}\ \mathbf{Fa}$$

which permits the union of any set of families to be called a family, while excluding from the family *of a given* D any Sort which is not one of the sequence of superstructs of this D.

As we have seen, endomorphisms are the only class of univalent but non-trivial functors which can be defined over any perfect Sort. Accordingly, with the exception of some cases reducible to those we include, all rational Sorts must be found in one or another family of endomorphisms built on one of the initial prime perfect Sorts whose Bip. Reps. were derived in Chapter VI. We have therefore only to consider which of these can be provided with representations.

7.C. FUNCTIONS OVER SUPERSTRUCTS

The definition of a superstruct Sort, and in the present context in particular a Sort of endomorphisms, does not allude to any particular functions which may be definable over it. In fact, I shall not seek to interpret any such function, in the course of the application of the theory in Part II. Nevertheless, Rule R2.3 of (3.2) requires us to represent all definable functors over such Sorts, other alongside than those which may enable the family to be continued further.

Let D_0 be a domain, with members x, y, ... , and let D_1 be its immediate superstruct, with members E, F,..., so that Ex, Fy, EFx, &c. denote members of $\bar{D}_0$, where EF denotes the successive application of first F and then E to x. But, of course, from Theorem 28, $\hat{=}$EF, so that = EFx FEx, and we can legitimately ask what function is denoted, in the formula EFx, by the null superfunctor whose arguments are E, F. Now if x is in the invariant subdomain of both E and F in D_0, then EFx = x; while if x is not in the invariant subdomain of E say, then Ex = k. It follows that the invariant subdomain of EF is the intersection of those of E, F separately.

We have seen that in general we can always define the functor J over any perfect Sort, by identifying JEF with $\hat{\cup}$EF. It is easy to see that this makes J identical to the above null superfunctor; for $\hat{\cup}$EFx = $\hat{\cup}$ExFx which is x if Ex = Fx = x but k otherwise. A formula for the representation of J over D_1 follows immediately from this:

$$\in,\mathrm{S}(\in\!\dot{!}\,\mathrm{D},\mathrm{D}(\in \mathbf{R}\,\mathrm{D})\,\mathrm{S})\,\hat{\cup}\,xy \Rightarrow\, = \mathbf{Rp}\mathrm{J}xy\ (\emptyset,\ \mathbf{Rp}x, \mathbf{Rp}y) \qquad (7.4.1)$$

which shows that J is to be represented by the null (super)functor in the appropriate context. An analogous representation for K is with the same protasis

$$\Rightarrow\, = \mathbf{Rp}\mathrm{K}xy(\emptyset,\ \mathbf{Rp}x,\ \mathbf{Rp},\mathrm{F}(=/,u(=\mathrm{F}uu)/x)) \qquad (7.4.2)$$

There is always, however, in such cases a less elegant method of representation, since we can find the set of representations invariant under the representation of a given functor, and carry out, in the biparitous representation algebra, operations on that.

This procedure, for example, would enable us to represent the functor U, defined so that the invariant subdomain of UEF shall be the *union* of those of E and F. Moreover, it is clear from Theorem (5.16) that all properties of functors over a given domain can be restated as properties of their invariant subdomains, and consequently any functor so defined could be represented in this way. It follows that there is no barrier to the representation of superstruct Sorts arising from difficulties with their functors. It remains to be seen whether we can in all cases find representations for their members.

7.D. THE CEDILLA FUNCTOR

There are two alternative ways of constructing endomorphism functors. One is to include a specification of the endomorphism in question as an argument under a special common functor; and the other is to define a functor so that, when given an appropriate complex on the substruct domain as a first argument, the required repertory of invariant subdomains should be generated in respect of the second arguments. The former method, being simpler, though necessitating the use of a classifying auxiliary functor, will be examined first.

All we need to do is to define a functor Ç such that, when extended by a symbol denoting a given invariant subdomain, it will note whether its second argument is or is not a member of the first, and gives the appropriate value (the argument itself or the smudge k), accordingly. Formally, the definition can be written:

$$\in \mathbf{R}_0\mathrm{D}, \quad \leqq \mathrm{D}d \Rightarrow\ = \text{Ç}d,x(\in dx)x, \quad = \text{Ç}d,x('\in dx)\mathbf{k}\mathrm{D}$$

(7.4a) Df. Ç

It must not be forgotten that since the superstruct Sort is perfect, the mapping from this D_1 onto the Ç-expressions is not reversible; that is to say, the notation *distinguishes* the elements of D_1 by the different symbols Çd, but does not thereby *identify* them. However, we may legitimately use such informal expressions as '**Im** Çd' *as if* to refer to elements of D_1 itself, provided no inferences are to be drawn from statements containing them. (This remark applies to all kinds of representations.) It is however true, and essential, that

$$\leqq \mathrm{D}_0 d, \quad \leqq \mathrm{D}_0 d', \quad = \mathbf{Im}\ \text{Ç}d\ \mathbf{Im}\ \text{Ç}d' \Leftrightarrow\ = dd' \tag{7.5}$$

where the superstruct domain D_1 is *not* mentioned in the protasis.

The possibility of using the definitions (7.4a) to construct a representation of D_1 depends entirely on the possibility of specifying the subdomains d of D_0. Since D_0 is perfect, this cannot be effected by the use of arbitrary letter symbols, such as the i, j &c. of Chapter VI, since their self-identity is valid only within a particular nexus, so that particular subdomains can never be associated with their members. But if D_0 has a Bip. Rep. H_0, then the values of d can be represented by explicit multiplets over H_0. In the case of $D2_0$ for example, the three Subsorts can be specified as /*i*, ˆ*ij* ˆ*ik*, using the representation $H2_0$ of (6.2). In the absence of a representation for D_0, the Ç-formulae do *not* provide a representation for any superstruct Sort.

Not every representation for D_0 can be used in this way. Let a, b be members of a domain D_0; then Ç ˆ*ab* represents the endomorphism on D_0 which has the Subsort ˆ*ab* as invariant, and the relation = *b* Ç ˆ*ab b* is unambiguously true because the *b*'s are concurrent (ˆ being declassified by Ç) and the symbols *a*, *b* are representatives of elements of D_0. But if we try to use formulae such as Ç ˆ*ab* as representatives in their turn of elements of D_1 in the hope of similarly defining representations in D_2 (that is, endomorphisms on D_1) we find that it does not work. Let G be the analogue of Ç at the next level; then an example of the kind of relation we shall come up with would be this:

$$= \text{Ç}\ \hat{}ab\ \text{G}\,\hat{3}\,\text{Ç}/a\ \text{Ç}\ \hat{}ab\ \text{Ç}/c\ \text{Ç}\ \hat{}ab \tag{7.6}$$

This equation is undecidable, because the various occurrences of *b* are non-concurrent, mainly on account of Theorem (2.24) which allows declassification to operate only once in a given formula. The functor Ç ˆ under G $\hat{3}$ is concurrent with the initial Ç ˆ, but Ç/ is different, and its *a* can be substituted by any twin under (4.2.2) without change in the truth-value of the expression.

To avoid this outcome, we must replace the 'search' functor Ç by a 'process' functor, operating algorithmically by steps reducible ultimately to functions defined over D_0. A representation in terms of cedilla functors can be validly applied to the endomorphisms on any Sort already provided with a representation *by simple multiplets*; but it is not itself such a representation, and does not admit of any further level of Bip. Reps. For the same reason, that is the non-identifiability of a multiplet such as ˆ*ap* in the expression Ç $\hat{3}$ ˆ*ap* ˆ*aq* ˆ*bq* ˆ*ap*, the cedilla functors can't be used to construct Bip. Reps of endomorphisms over derived Sorts represented in terms of their initial prime Sorts (assuming these to be rational); this

I mentioned in 7.A. Nor are there cedilla representations for Sorts represented by complexes other than simple multiplets, for the same reason.

This conclusion prompts us to distinguish *medial* from *terminal* superstructs, thus:

$$\in \mathbf{Fa}\, D_0\, D_r \Rightarrow \in \mathbf{Me}\,, D_r(\nparallel \mathbf{En}\, D_r \emptyset) \qquad (7.7.1)\ \text{Df. } \mathbf{Me}$$

$$\in \mathbf{Te}\,, D_r(= \mathbf{En}\, D_r \emptyset) \qquad (7.7.2)\ \text{Df. } \mathbf{Te}$$

There are actually *no* endomorphisms on a terminal Sort, because if its only representation is by cedillas or higher complexes there is no way of defining them except by process functors which ((7.7.2) assumes) are not available. In some cases there may be two representations available for a given Sort of endomorphisms, one by cedillas, making it terminal, and another allowing the family to be continued. I take the view in such cases that we have here two distinct Sorts, because they differ in what functions are definable on them, so that they warrant separate correlations in the physical plane.

7.E. THE HEX FUNCTOR ⟊

We have found that the inclusion of a specification of the invariant subdomain as an argument under the cedilla functor to represent a given endomorphism leads to a breakdown of concurrence among the symbols identifying particular members of the substruct domain. There is, however, an alternative procedure, namely to define a functor which specifies an algorithm deriving the invariant subdomain from some complex on the substruct domain by its manner of dependence on the possible second arguments denoting individual members of the substruct domain.

Let D_0 be a rational Sort, and D_1 its immediate superstruct; let them have Bip. Reps H_0, H_1 with members h_0, h_1 respectively. We shall seek to find a common functor ⟊ such that for any given h_0, h_1 the formula ⟊$h_1 h_0$ either equals h_0 or not. (Strictly speaking, the required function would be K_0⟊$h_1 h_0 h_0$ where K_0 is the K-type functor over H_0, but to simplify matters I shall omit the symbol K_0.) We can now accept the subformula ⟊h_1 as representing that endomorphism over D_0 which has the same invariant subdomain. No classifying functors are involved, and no trouble arises through non-concurrence, providing that ⟊ can be appropriately defined and a suitable h_1 exists. To satisfy R1.1 and R2.3 all these definitions must be ultimately translatable into the terms of the initial Sort of the family.

Those values of h_0 which are invariant under a given ⟊h_1 have an obvious analogy to the eigenvectors under a given matrix; this suggests that the complexes used for h_1 should be analogous to arrays of elements of the form of h_0, the functor $)\!($ itself being the counterpart of the null functor denoting matrix multiplication. This analogy provides a model which is successful in leading to a solution of our problem.

We must define $)\!($, which I read as *hex*, so as to be applicable to the representation of any domain. As a novel functor, this involves at least one arbitrary innovation, which (from the limited variety of functors over triparitous Sorts) must be made in the Bip. Reps.; and to ensure that everything is referrable to the initial domain D_0 this must be done in H_0 and once only. The 'arbitrary innovation' takes the form of an auxiliary functor X (called *hox*) closed over H_0. Besides X the definition of $)\!($ can make use of any of the functors constitutive of the initial domain D_0, but no others. To waive this restriction would enable us to redefine $)\!($ for each H_n, but such redefinitions would not then be referrable to D_0, so infringing R2.3. The same rule also requires that at the initial level the definition of $)\!($ must be in terms of D_0 rather than of H_0, so that every level can be ultimately expanded into a form explicitly based on D_0 itself.

The essential constraints on $)\!($ are therefore

$$n > 0 \Rightarrow \in H_n)\!(,h_{n+1} (\in H_{n+1} h_{n+1}),h_n(\in H_n h_n) \qquad (7.8.1)$$

$$(\,) \quad \Rightarrow \in \bar{D}_0)\!(,h_1 (\in H_1 h_1),x(\in D_0\, x) \qquad (7.8.2)$$

The problem now before us is to decide what kind of complexes we can use for the various h_n. We have seen that there is a representation, valid by (3.2), for each initial Sort Dn_0 by n-plets, up to $n = 3$. This suggests that we should look elsewhere than among simple multiplets for the members of what must evidently be a new class of Sorts. The simplest of such would evidently be, in the case of $D2_0$, pairs-of-pairs; thus the complexes forming the representation $H2_1$ of $D2_1$ would be of the form $\hat{}\hat{}h_1 h_2\, \hat{}h_3 h_4$, where $\in \bar{D}2_0\, h_i$ for all i. The formula $)\!(\, \hat{}\hat{}h_1 h_2\, \hat{}h_3 h_4\, x$ must then denote a member of $\bar{D}2_0$ represented by one of $H2_0$, that is, by one of the pairs $/i$, $\hat{}ij$, $\hat{}ik$.

The definition of $)\!($ is constrained not only by (7.8), but also by the requirement for total symmetry among the h's, and for X to operate directly on **Rp**x, so that any formula $\mathsf{X}xy$ directly represents a member of $\bar{D}2_0$. We are therefore reduced to the following:

$$\in\ ,D(=\mathbf{Rp}\,D,H(\in H\,\hat{4}\,h_1 h_2 h_3 h_4))x \Rightarrow$$

$$=\mathbf{Rp})\!(\,\hat{}\hat{}h_1 h_2\hat{}h_3 h_4 x \quad \hat{F}\ \mathsf{X}\,\hat{}h_1 h_2\,\mathbf{Rp}\,x\ \mathsf{X}\,\hat{}h_3 h_4\,\mathbf{Rp}\,x$$

(7.9) Df. $)\!($

where $\hat{\mathrm{F}}$ stands for either $\hat{\mathrm{G}}$ or $\hat{\mathrm{J}}$. We have still to define $\mathbf{X}$ and to choose a value for the $\hat{\mathrm{F}}$ of (7.9).

7.F. THE AUXILIARY FUNCTOR $\mathbf{X}$

Since $H2_0$ is required to be an ordinable Sort, we can define any functor over it by means of an operational table, since the objections to this style of definition do not apply for such Sorts. $\mathbf{X}$ must not reduce to either $\hat{\mathrm{G}}$ or $\hat{\mathrm{J}}$. It must be commutative, since its arguments represent twins, and likewise idempotent, and to maximize the chances of arriving at invariant subdomains the various products should be all equally frequent. Associativity is not required, since the occasion for its relevance will not arise. Any function consonant with these requirements must have a table such as (7.10) or some simple commutant of it:

$\mathbf{X}$	$\hat{}ii$	$\hat{}ij$	$\hat{}ik$
$\hat{}ii$	$\hat{}ii$	$\hat{}ii$	$\hat{}ik$
$\hat{}ij$	$\hat{}ii$	$\hat{}ij$	$\hat{}ij$
$\hat{}ik$	$\hat{}ik$	$\hat{}ij$	$\hat{}ik$

(7.10) Df. $\mathbf{X}$

It will be observed that, though $\mathbf{X}$ is defined over $H2_0$, we could equally define a functor X as if over $\bar{D}2_0$ by the formula $= \mathbf{X}\,\mathbf{Rp}\,x\,\mathbf{Rp}\,y\ \hat{}i\,\mathrm{X}\,xy$; using this, the apodosis of (7.9) becomes $\hat{\mathrm{F}}\ \hat{}i\mathrm{X}h_2x\ \hat{}i\mathrm{X}h_4x$ which is equal to $\hat{}i\,\hat{\mathrm{F}}\,\mathrm{X}h_2x\mathrm{X}h_4x$, and this represents simply $\hat{\mathrm{F}}\,\mathrm{X}h_2x\,\mathrm{X}h_4x$, so that (7.9) can be rewritten as

$$\in{,}\mathrm{D}(=\mathbf{Rp}\,\mathrm{D},\mathrm{H}(\in\mathrm{H}h_i))\,x \Rightarrow\ = \not{\mathrm{X}}\ \hat{}\hat{}ih_2\ \hat{}ih_4\,x\,\hat{\mathrm{F}}\ \mathrm{X}h_2x\ \mathrm{X}h_4x$$

(7.9a) Df. $\not{\mathrm{X}}$

in which statement all the three apodosis x's are concurrent with the protasis x, while the h's (including i) derive from those of (7.9) which are defined as members of the ordinable Sort H and so are immune from the concurrence rules. Thus (7.9) and (7.9a) together with the reduced X-form of (7.10) ensure the complete reducibility of all $\not{\mathrm{X}}$-expressions to formulae in $\bar{D}2_0$, as required by R2.3. Note however that X is not a genuine functor over $\bar{D}2_0$, because it is not associative ($\ddagger$ XXijk XiXjk). This, however, does not affect (7.9a), since this statement has only to be applied for values of x in the sharp domain $D2_0$ excluding k, and over this domain X is associative and therefore acceptable as a functor.

The only point left undivided is the value of F̂. We shall find that the accumulation of blurs is the major factor in reducing the variety of invariant subdomains which any level of ⚹-representation can attain; and therefore I choose Ĝ as giving fewer blurs than Ĵ does. Over the six distinct members of $H2_1$ and the three of $\bar{D}2_0$ or $H2_0$ (of which *k* is shown for completeness only), the table for ⚹ as defined by (7.9, 9a, & 10) is:

	$\in \bar{D}2_0$:	*i*	*j*	*k*	Invariant Subdomain of $D2_0$
	$\in H2_0$:	ˆ*ii* = *m*	ˆ*ij* = *n*	ˆ*ik* = *p*	
∈ $H2^2$ F		⚹F*m*	⚹F*n*	⚹F*p*	
M =	ˆˆ*ii* ˆ*ii*	***i***	*i*	*i*	/*i*
N =	ˆˆ*ii* ˆ*ij*	***i***	***j***	*k*	ˆ*ij*
P =	ˆˆ*ii* ˆ*ik*	*k*	***j***	*i*	/*j*
Q =	ˆˆ*ij* ˆ*ij*	***i***	*i*	*i*	/*i*
R =	ˆˆ*ij* ˆ*ik*	*k*	*i*	*k*	∅
S =	ˆˆ*ik* ˆ*ik*	***i***	*i*	*i*	/*i*

(7.11)

As an example of the calculations involved in this table, take ⚹P*j*, in full ⚹ ˆˆ*ii* ˆ*ik j* = ĜX*ij*X*kj* = Ĝ*ij* = *j* as given. Again, ⚹R*i* = ⚹ ˆˆ*ij* ˆ*ik i* = ĜX*ji*X*ki* = Ĝ*ik* = *k* also as given.

It appears from (7.11) that our choice for $H2_1$ lies between three Subsorts of $H2^2$, namely $\hat{3}$MNP, $\hat{3}$NPQ, and $\hat{3}$NPS.

7.G. PAIR TREES

Our next step is to find a set of seven complexes which, under an extended definition of the Hex functor, will have for their invariant subdomains each of the subsets of one of the above sets of three; such a set of seven will then serve as representative of the members of $D2_2$, understood as endomorphisms over $D2_1$. The model presented by the first step in the representation of the family D2 suggests that we shall need to explore, for this second step, a class of complexes more elaborate than the pairs-of-pairs $H2^2$ used above. For this purpose, I now introduce the notion of *pair-trees*, comprising, besides pairs of pairs, pairs of any other isomorphic members of the class; these I define as the class **H2̂D**, as follows:

∈**S**$\hat{\cup}$DH ⇒ = **H**2̂D/ ˆ̂H, H(∈$\hat{\cup}$/D**H2̂**DH)   (7.12) Df. **H2**

I shall call the Sort D as before the *initial* Sort of the pair-trees $\mathbf{H\hat{2}D}$, and I shall assign to it the *rank* 0, while if a Sort of pair-trees H has the rank r, then $\hat{\ }$HH has the rank $r + 1$. Besides the unordered pair-trees defined in (7.12), we can also use *ordered* pair-trees, denoted by $\mathbf{H\ddot{2}D}$, since in general for $r > 0$ the elements to be paired will be bipar. An ordered pair-tree of rank r will contain a total of 2^r elements of D in its make-up, and this number I call the *extent* of the pair-tree; the Sort of all pair-trees of rank r on a given initial D I shall denote by $H\hat{2}^r$ or $H\ddot{2}^r$, omitting D if this is $D2_0$. The least number of non-identical symbols needed to express a given pair-tree may be called its *complexity*; for example, ^^*ii* ^*ij* has an extent of 4, but since three of the elements are *i* it can be written ^/*i* ^*ij* so its complexity is 3. The sequence of multiplet functors /, ^&c. obtained by deleting all their arguments (in the above case, ^/ ^) is the *shape* of the pair-tree; while its *sharpness* is related inversely to the number and range of component symbols denoting blurs in D. The shape and sharpness of the six members of $H\hat{2}^2$ (7.11) are as follows: /, 1; ^/ ^, 1; ^/ ^, 2; / ^, 1; ^^^, 2; and / ^, 4 respectively. Pair-trees over a perfect initial which differ in either shape or sharpness are distinct.

Every Sort of pair-trees is a mixed Sort. Since twins cannot be ordered, the ordinal of a mixed Sort differs from its cardinal; and since for Bip. Reps we need to have ordinable Subsorts only, the proper measure of the size of a Sort of pair-trees for our purpose is its ordinal, not its cardinal. Writing ^{0}H for the ordinal of a Sort H and Δx for $x(x + 1)/2$, the general formulae for ordered and unordered pair-trees of rank r over a perfect initial D are:

$$\in \mathbf{R}\,D,\quad \in \mathbf{I}\,r,\quad \in \mathbf{H\hat{2}D}\; H\hat{2}^r \Rightarrow = {}^0H\hat{2}^r\; \Delta^r\, {}^cD \tag{7.13.1}$$

$$\in \mathbf{H\ddot{2}D}\; H\ddot{2}^r \Rightarrow = {}^0H\ddot{2}^r\; (\Delta^2\, {}^cD)^{2^{r-2}} \tag{7.13.2}$$

Note that for ordered pair-trees, the ordering only begins at the third level, since not only the simple unordered pairs of $H2_0$, but the pairs of these forming $H2_1$ are equivalent in either order when functioning as independent pair-trees; if however order of pairs is allowed to be significant, $\dot{-}$'' ^*ii* ^*ij* '' ^*ij* ^*ii*, so that as operands under further Hex functors they must be distinguished from the third level onwards. This gives the following table of the numbers of distinct pair-trees:

Rank	r	:	0	1	2	3	4	5	6	…	8	…	10	
Extent	2^r	:	1	2	4	8	16	32	64	…	256	…	1024	
Ordinal	$^0H\hat{2}^r$	:	2	3	6	21	231	$^3 10^4$	$^3 10^8$	…	$^1 10^{33}$	…	$^1 10^{132}$	(7.13a)
	$^0H\ddot{2}^r$	:	2	3	6	36	1296	$^2 10^6$	$^1 10^{12}$	…	$^2 10^{48}$	…	$^2 10^{193}$	
Cardinal	$^cD2_{r/2}$	:	2	–	3	–	7	–	127	…	$^1 10^{38}$	…	$^1 10^{10^{38}}$	

The last row shows the numbers of elements in the successive domains which have to be represented. With ordered pair-trees we cannot go beyond $D2_4$, while with unordered ones the limit is $D2_3$.

7.H. GENERAL DEFINITION OF $\between$ OVER PAIR-TREES

Up to now I have defined $\between$ only for arguments in $H\hat{2}_0^2$ and $H2_0$ respectively. It is clearly necessary, if we are to construct representations for Sorts superstruct to $D2_1$ to extend this definition to higher ranks of pair-trees. The definition must be recursive, and must terminate in (7.9) or an equivalent. Since the last step must end in $D2_0$ itself, there is a difference of two in the formal ranks of the arguments in (7.9a), and this feature must recur at each stage, so we can use only even-rank pair-trees in our representations.

To make the procedure clear, I shall write any pair-tree of rank r as ${}^rh$ or ${}^rg$, where $= {}^rh \, '' \, {}^{r-1}h_1 \, {}^{r-1}h_2$. Each stage of the recursion must bring in two new $''$ pairings to effect the required rank reductions, and the functor $\hat{\mathrm{F}}$ of (7.9) must also reappear once at each stage; I shall therefore represent each stage as three steps. $\hat{\mathrm{F}}$ may be either $\hat{\mathrm{J}}$ or $\hat{\mathrm{G}}$ (which, existing for $D2_0$, can be simply extended to any *ordered* pair-trees on $D2_0$); of the two, $\hat{\mathrm{G}}$ is preferable to $\hat{\mathrm{J}}$ as giving fewer blurred values, and so allowing for a greater variety of invariants.

If the $\hat{\mathrm{G}}$ functor is introduced in expanding a $\between$-formula with even ranks of pair-trees, $\between {}^4h \, {}^2g$ will reduce to a pair-of-pairs of $\hat{\mathrm{G}}$-formulae on $\between {}^2h \, {}^0g$; but ${}^0g$ is outside the representation, being a member of $D2_0$. It follows that the $\hat{\mathrm{G}}$ must appear in explication of an odd-rank formula. The only consistent and non-degenerate set of formulae for defining $\between$ is thus as follows:

$$r = 2i,\ i > 1 \Rightarrow\ = \between {}^{r+2}h \, {}^rg \ '' \between {}^{r+1}h_1^r \, g \between {}^{r+1}h_2^r \, g \tag{7.14.1}$$

$$= \between {}^{r+1}h_1 \, {}^rg \, \hat{\mathrm{G}} \between {}^rh_{11} \, {}^{r-1}g_1 \between {}^rh_{12} \, {}^{r-1}g_2 \tag{7.14.2}$$

$$= \between {}^rh_{11} \, {}^{r-1}g_1 \ '' \between {}^rh_{11} \, {}^{r-2}g_{11} \between {}^rh_{11} \, {}^{r-2}g_{12} \tag{7.14.3a}$$

but

$$r = 2 \qquad \Rightarrow\ = \between {}^2h_{11} \, {}^1g_1 \, \hat{\mathrm{G}} \, \mathbf{X} \, {}^1h_{111} \, {}^1g_1 \ \mathbf{X} \, {}^1h_{112} \, {}^1g_1 \tag{7.14.3b) Df. $\between$}$$

The last statement terminates the recursion, and restates (7.9) in the notation used here. The first term can be written $\between {}^2h_{11}$ **Rp** x which clearly represents $\divideontimes\,{}^2h_{11}\,x$ (where $\in D2_0\,x$), the form given in (7.9). Note that here also $\divideontimes$ is replaced at the end by **X**, as defined in (7.10).

7.J. REPRESENTATION OF $\bar{D}2_2$ &C., BY HEX FORMULAE

To show that this does work we have to find a set of seven rank-4 pair-trees from $H\ddot{2}^4$, which, with $\divideontimes$ as defined in (7.14), will give distinct subsorts of $D2_1$ (already represented in (7.11) by one of 3 sets of members of $H\hat{2}^2$ constituting the representation of $H2_1$) as their invariant subdomains. Since there are 1296 distinct pair-trees in $H\ddot{2}^4$ the search will be more laborious than it was for $H2_1$.

The procedure is simplified by the fact that the work of (7.14.2) is simple enough to be done in one's head; and, as this is the last step of the recursion before it terminates, we shall not need (7.14.3a). All we need is to tabulate the results of (7.14.2) for each of the 21 pair-trees in $H\hat{2}^3$ (we can still identify commutants in $H\hat{2}^2$ for this purpose so we don't need 36), form their $\hat{G}$-products two by two, and then pick out the invariants.

Much depends on the set of pair-trees we choose for the representation $H2_1$. The choice lies between $\hat{3}$MNP, $\hat{3}$NPQ, and $\hat{3}$NPS; the first of these is not possible, because no member of $H\ddot{2}^4$ has the Subsort ˆMN or ˆMP invariant, but either of the others can be accommodated. I shall perform the calculations as for $\hat{3}$NPQ; but if the resulting choices for $H2_2$, of which there are a wide range, are all excluded as invariant under members of $H\ddot{2}^6$, which is hardly likely, we might try again with $\hat{3}$NPS. As a reminder, note that

$$\mathrm{N} = \hat{}mn \quad \mathrm{P} = \hat{}mp \quad \mathrm{Q} = \hat{}nn \tag{7.14a}$$

Though not in fact needed till $H\ddot{2}^6$, the definition of $\hat{G}$ over '' is

$$\in \mathbf{Q}\,\hat{\cup}\,pqrs \Rightarrow\, = \hat{G}\,''pq\,''rs\,''\hat{G}pr\hat{G}qs \tag{7.14b}$$

The columns headed N*, P*, R* give the results with N, P, R and the *commutant* of the pair ''EF shown on the left; Thus $\divideontimes$''PQ N = p (row j col. N) whereas $\divideontimes$''QP N = n (row j col N*). The invariants shown in these columns at the bottom are those ''EF pairs whose commutants make N, P, or Q invariant; thus we find d, h for P*, meaning that P is invariant under '' ''RM ''RN. Further invariants are found from one column under P and the same under P*: thus 1 on the left under P with g on the left under P* show $\divideontimes$ '' ''RR ''QN P = P.

We can pick out from this table a variety of possible sets of members of $H\ddot{2}^4$ which could serve as representations for $D2_2$; thus with the set

		Values of ⁎″EF X where $\in H\hat{2}^2$ X, and X =								
$\in H2^3$ ″EF		M	N	P	Q	R	S	N*	P*	R*
a	″MM	*m*	*m*	*m*	*m*	*m*	*m*	*m*	*m*	*m*
b	″MN	*m*	*n*	*p*	*n*	*p*	*p*	*m*	*m*	*n*
c	″MP	*p*	*n*	*m*	*n*	*m*	*m*	*p*	*p*	*n*
d	″MR	*p*	*m*	*p*	*m*	*p*	*p*	*p*	*p*	*m*
e	″NN	*m*	*n*	*p*	*m*	*p*	*m*	*n*	*p*	*p*
f	″NP	*p*	*n*	*m*	*m*	*n*	*p*	*p*	*m*	*p*
g	″NQ	*m*	*m*	*m*	*n*	*n*	*p*	*n*	*p*	*p*
h	″NR	*p*	*m*	*p*	*n*	*p*	*m*	*p*	*m*	*p*
i	″PP	*m*	*p*	*p*	*m*	*n*	*m*	*p*	*p*	*n*
j	″PQ	*p*	*p*	*p*	*n*	*n*	*m*	*n*	*m*	*m*
k	″PR	*m*	*p*	*m*	*n*	*p*	*p*	*p*	*p*	*m*
l	″RR	*m*	*p*	*m*	*m*	*p*	*m*		*m*	*p*
m	″RS	*p*	*p*	*p*	*m*	*p*	*p*	*m*	*p*	*p*
X =		″*mm*	″*mn*	″*mp*	″*nn*	″*np*	″*pp*	″*nm*	″*pm*	″*pn*
pick 1 ″EF		a a	a b	a b	b b	f b	b b	e a	c a	e a
from each		b b	d c	c d	c c	g d	d d	g b	d b	f b
column to		e e	g e	f e	g g	i h	f f	j m	e f	g i
get a pair-		g g	h f	g h	h h	j k	g g		g h	h
tree with X		i i		k j	j j	l	k k		i j	l
as invariant		k k		l m	k k	m	m m		k l	m
		l l							m	

$\hat{3}$ NPQ as $H2_1$, we have

pair-trees: ″″NQ″MN ″″MR″NN ″″NR″MN ″″PR″PQ
invariants: N, P, Q N, P N, Q P, Q

pair-trees: ″″MM″NP ″″RR″RS ″″MN″PQ
invariants: N P Q (7.16.1)

while, with the alternative $H2_1$ namely $\hat{3}$ NPS we could use

pair-trees: ″″NQ″MN ″″MM″MN ″″NQ″NP ″″PR″RS
invariants: N, P, S N, P N, S P, S

pair-trees: ″″NR″NN ″″MM″MR ″″RS″RM
invariants: N P S (7.16.2)

There is, however, no way to make ˆMN or ˆMP an invariant subdomain of $D2_2$, whatever combinations we try to build up a pair-tree; it follows that we cannot use the set $\hat{3}$MNP as our value of $H2_1$ to represent $D2_1$ if the family D2 is to be continued. Similar dead-ends doubtless occur at every level.

The next step is now to find a set of 127 members of $H\ddot{2}^6$, out of the 10^{12} available, which will have each and every non-empty Subset of at least one $H2_2$ as their invariant subdomains. Though I have given above only one example of a $H2_2$ for each of two $H2_1$, both of these have a great number of alternatives available, and in view of this and of the much greater number of pair-trees available for search in $H\ddot{2}^6$, it seems very unlikely that the search will fail. Actually to calculate the particular pair-trees forming even one set of 127 is a task calling for a computer, and there is no other way of proving that $H2_3$ exists, unless someone can devise an analytic proof. After that we have to find $17_{10}37$ pair-trees out of the $2_{10}48$ in $H\ddot{2}^8$ to fill the role of $H2_4$. Once again there is no good ground to expect failure, but proof of existence obviously depends solely on finding an analytical method; even to print out the list would take longer than the age of the universe.

It will be seen from (7.13a) that the next level $H2_5$ needs to have $1_{10}10^{36}$ members. Since there are in $H\ddot{2}^{10}$ only $2_{10}193$ pair-trees, there is no need to look further; there is obviously no representation available for this Sort, unless a completely new method of construction can be devised. I have no proof to offer that this is impossible; but it is likely that if such a method exists, it will take the form of a loosening of one or another of the restrictions which I have, perhaps arbitrarily, imposed on my 'pair-trees', and I *can* show, by elimination, that this route at least is closed.

7.K. ARE THERE ALTERNATIVES TO HEX?

Most precedents would suggest that the method of exhaustion is unlikely to be successful in so open-ended a quest as that of finding alternatives to the Hex functor used in the above procedure; but the relative poverty of triparitous theory here comes to our aid, and we can lop off the branches of the choice-tree in a couple of pages. In fact, alternatives to pair-trees can only come from pair-Sorts or power-Sorts. The latter are incorporated in the cedilla notation, and their use entails specifying classes in a way which we have found to be non-iterable; the former include the Sort-powers from which any plausible improvement on pair-trees must come, for if disparate pair-Sorts are made to represent endomorphisms, they will turn out to be Sort-powers in disguise. As I have already taken care in defining my pair-trees to select (wherever a choice arises) the most productive alternative, the search for improvements in the scheme narrows to two possibilities: to use more widely-spaced ranks of pair-trees, and to construct triplet- or multiplet-trees instead of mere pairs.

Before looking into these possibilities, I should perhaps direct attention to an apparent anomaly in the definition (7.14), where we have a regular alternation of '' -steps and $\hat{G}$- steps until the terminal step, which is another $\hat{G}$-step. This is actually an artifact of the notation, in that since the value of a X-formula is a pair on $D2_0$ and since G distributes over ^, (7.14.3b) can be represented as two steps, introducing ^ and G in the same order as the preceding steps. Nevertheless, it alerts us to the possibility of inverting the order of using $\hat{G}$ and '' in all previous steps. But it's no use; the effect is to arrest the construction at $H2_2$, as might be expected from what is really an arbitrary violation of symmetry.

What then would we get if we were to choose representations from pair-trees of ranks more widely spaced? Suppose, for instance, we ignored $H\ddot{2}^6$ and sought for representations of $D2_3$ in $H\ddot{2}^8$ instead. This will require a modification of (7.14) where $\hat{G}$ appears in two non-contiguous steps; for if not, we must have two pairings introduced either before or after the $\hat{G}$-step, which will fail to give a X-product in $H\hat{2}^4$ as the proposal requires. But if we have two $\hat{G}$-steps alternating with ''-steps, we shall necessarily have an intermediate, supposedly uninterpreted, pair-tree belonging to $H\ddot{2}^6$; let us call this intermediate X. We can then write for any H* in $H\ddot{2}^8$

$$\in H\ddot{2}^8 \,\hat{\cup}\, H{*}H,\quad \in H\ddot{2}^6 X,\quad \in H\ddot{2}^4 h \Rightarrow {=}\, H{*}h\, HXh \qquad (7.17.1)$$

But, with the same protasis P together with

$$P,\quad {=}\, h\, H,X({=}\, XH,X({=}\, Xhh))h \Rightarrow {=}\, H{*}h\,h \qquad (7.17.2a)$$

for $= XHX \Rightarrow = H,X(= XHX)h\, Xh$ and so $= Xhh \Rightarrow = H,X(= XH,X(= Xhh))h\,h$. But it is equally true that

$$P,\quad {=}\, H{*}h\,h \Rightarrow {=}\, H,X({=}\, XH,X({=}\, Xhh))h \qquad (7.17.2b)$$

for to negate the first = implies $= H{*}h\mathbf{Rp}k$, while $\nmid XHX \Rightarrow = HXH\ddot{2}^8 \Rightarrow$

$= \mathrm{H}^{*}h\,\mathbf{Rp}k$, and ǂ $\mathrm{X}hh \Rightarrow = \mathrm{H}^{*}h\,\mathbf{Rp}k$, all these from Theorem (5.17). Therefore (7.17.2) is an equivalence, which entails that every invariance relation under H* can be reduced to one under an X invariant under some other H. Thus, the existence of a representation in $\mathrm{H}\ddot{2}^{8}$ requires that there be also one in $\mathrm{H}\ddot{2}^{6}$, and nothing has been gained by skipping a rank.

The only remaining expedient would be to substitute complexes other than pairs in our constructions, and so produce multiplet-trees. To take the simplest case, triplets, we would have in place of (7.13.2) that $= {}^{0}\mathrm{H}\ddot{3}^{r}$ $\Delta^{2}\,{}^{c}\mathrm{D})^{3^{r-2}}$, which, for $r = 10$, gives $= {}^{0}\mathrm{H}\ddot{3}^{10}\,1_{10}\,5105$; but even so 5105 is vastly less than 10^{38}, and it is evident that if we are to break the barrier into $\mathrm{D2}_{5}$ we shall need to use very much greater multiplets than triplets. And calculation shows that even triplets, applied to $\mathrm{D2}_{0}$, produce nothing but redundant components; though the method might have an application to the family ′D3, which I shall now examine.

7.L. REPRESENTATION OF THE FAMILY D3

In fact, it is unnecessary to consider whether a representation of D3 by triplet-trees is feasible or not, because apart from $\mathrm{D3}_{0}$ which has already been found to have a representation by triplets, we can show that all the other members of this family have already been dealt with. $\bar{\mathrm{D}}\mathrm{3}_{0}$ has all three functors J, H, K; this would seem to distinguish it from the equipotent $\mathrm{D2}_{1}$ for which I have mentioned no functors definable over its representation $\mathrm{H2}_{1}$. However, there is an alternative representation for this Sort, which I call $\mathrm{H2}_{1}'$, for which J and K (and therefore also H) have representatives. This consists in identifying the members of the representation ${}_{3}\hat{}\mathrm{NPQ}$ as *ordered* pairs, thus:

$$= \mathrm{N}'\,''/i\,\hat{}ij; \quad = \mathrm{P}'\,''/i\,\hat{}ik; \quad = \mathrm{Q}'/\,\hat{}ij \tag{7.18}$$

in relation to which we can define $\hat{\mathrm{J}}$, $\hat{\mathrm{K}}$ representing J, K over $\mathrm{H2}_{1}'$ as follows:

$$= \hat{\mathrm{J}}\,''pq\,''rs\;''\hat{\mathrm{G}}\,\hat{}ij\,\mathrm{X}\hat{\mathrm{G}}p\,\hat{}ij\,\hat{\mathrm{G}}r\,\hat{}ij\;\;\hat{\mathrm{G}}\,\hat{}ij\,\mathrm{X}\hat{\mathrm{G}}q\,\hat{}ij\,\hat{\mathrm{G}}s\,\hat{}ij \tag{7.19.1}$$

$$* = [= \hat{\mathrm{K}}\,''pq\,''rs\,\hat{\mathrm{J}}\,''pq\,''rs]\,''pq\,''rs, \quad = \hat{K}\,''\mathrm{pq}\,''rs\,''\,\hat{}ij\,\hat{}ik \tag{7.19.2}$$

where in the first term of (7.19.2) I use the conditional notation of 2.J.

These definitions lead to the following operational tables over the members N′, P*, Q′ of $\mathrm{H2}_{1}'$ (the primes omitted):

$\hat{J}$	N	P	Q
N	N	P	Q
P	P	P	R
Q	Q	R	Q

$\hat{K}$	N	P	Q
N	N	R	R
P	R	P	R
Q	R	R	Q

(7.19a)

The table for $\hat{K}$ is (as usual) trivial, but that for $\hat{J}$ will be found to accord with the requirement that the invariant subdomain of $\hat{J}$ X Y is the intersection of those of X and Y (that of R being, as shown in (7.11), empty).

It follows from this excursus that $D3_0$ and $D2_1$ both have representations under which all the functors of the former can be defined also for the latter, from which we conclude that they are in effect identical Sorts; though the fact that $D3_0$ is prime and $D2_1$ is derived (at least formally) will be understood to entitle them to separate physical correlations in Part II. However, even this distinction disappears at the next level, and $D3_1$ is identical with $D2_2$, apart front its family connections. All members of D3 have therefore representations isomorphic with those for members of D2.

7.M. REPRESENTATION OF THE FAMILY D∞

Infinite Sorts also have their own superstructs, and while these, at least under the treatment C of Chapter VI, each have as many members as their immediate substructs have subsorts, generating a sequence of transfinite cardinals with strictly no limit, it is possible as with the initial member $D\infty_0$ to represent only a denumerable Subsort of each; this is 'treatment B' of 6.K. Under treatment A, which refuses to consider the use of completed infinities, it is impossible to specify any endomorphism on an infinite Sort (because this requires us to discover for *every* element whether it is invariant), the family D∞ reduces to $D\infty_0$ alone.

Pursuing the more interesting treatment B, the representation of a denumerable Subsort of $D\infty_1$ can be effected by cedilla functors. This cannot be carried further as it stands; but since it is only a denumerable Subsort, that is to say, isomorphic with $H\infty_0$, we can correlate the cedilla functors with unbounded multiplets, as used for $D\infty_0$, and using these construct a cedilla representation for $D\infty_2$, and so on. (This will not work for finite Sorts (except $D2_1/D3_0$) because we have no representation by simple multiplets for the 7 members of $D2_2$.) The various representations in this series can be distinguished from each other by incorporating some previously unused member of $D\infty_0$ in that for each $D\infty_n$.

Under treatment C, every $D\infty_n$ can in principle be represented by an equipotent class of transfinite multiplets. I do not take this very seriously.

Reverting to treatment B we note that, as for the other infinite family D1, all the Sorts as actually represented are identical, except for their place in the series; and in both cases the union of all the Sorts is infinite. Such a union could be presented as containing every other representable Sort as a Subsort; that is,

$$\leqq, D(\in \mathbf{S}\, D), X(\nmid \emptyset/, H(= H\, \mathbf{Rp} X)) \Leftrightarrow = D\,!\,D \qquad (7.20)\ \text{Df. } D\,!$$

7.N. RATIONAL SORTS

This concludes the main part of our search, that is for biparitous representations of as many initial and superstruct Sorts as have them. Our ultimate object, however, is to list all the *rational* Sorts. These have been defined as having representations in terms of *their own* members, whereas most of our representations have been (reducible to) formulae over *initial* Sorts which are not themselves part of the Sort being represented. Only the initial Sorts themselves have thus been shown to qualify as 'rational'. For example, $H2_1$ consists of pairs of pairs of elements of $D2_0$: but is claimed to be a representation not of $D2_0$ but of $D2_1$. However, the union $\cup\, H2_0\, H2_1$ certainly is a representation of $\cup\, D2_0\, D2_1$, which includes the elements (of $D2_0$) in terms of which *all* the members of $\cup H2_0\, H2_1$ are defined, so that we can claim $\cup D2_0\, D2_1$ as a truly rational Sort, satisfying the rules given in (3.2).

In general, then, the rational Sorts we have found are the unions of initial subsequences of the families of endomorphisms, provided all the constituent Sorts are severally representable in the way defined in this chapter. I shall denote such a union by its last term underlined: thus, $\underline{D2_2}$ stands for $\cup \cup D2_0\, D2_1\, D2_2$. Terminal Sorts are distinguished by an asterisk, as in $\underline{D2_4^*}$; apart from this example, all terminal Sorts are so because they have cedilla representations.

The identification of rational Sorts with *unions* of perfect Sorts raises a question in relation to the finding in 4.K that such unions can be defined (except 'to within twinship' – meaning that $\cup XY$ includes anything that is *twin to* any member of X or Y besides their proper members) only between Subsorts of an independently defined Sort. The solution though necessary is simple: since, by definition, all our perfect Sorts do have representations, that Sort which is represented by the union of all the representations (which being biparitous are unionable) is an independently defined Sort of which all the others are Subsorts. It is the D! of (7.20). It is not necessary that this Sort be itself representable other than by a union (in fact, under none of the 'treatments' of infinite Sorts does D! *both* exist *and* have a representation); but that the well-definedness of any collection of rational Sorts should depend on the existence of a whole of which they are parts, even in the limiting case of treatment A where = D!D∞_0, is perhaps of some philosophical interest.

7.O. CATALOGUE OF RATIONAL SORTS

The following Rational Sorts have been shown to exist (except $\underline{D2}_3$ and $\underline{D2}^*_4$ whose representations have not been constructed and have not been proved to exist, though the contrary is very unlikely). Apart from pair-Sorts between RS's, there are no other Rational Sorts besides those listed below:

Sort	Cardinal		Functions	Symbol H	(All Definitions of Representations H)
$D0_0$	0		–	$H0_0$	$H0_0 = \emptyset$
$D1_0$	1		I	$H1_0$	$H1_0 = /, I(=I, x(\in D1_0\, x)x)$
$\underline{D1}_n$	n		$I_{\cup \cdots \cup}$	$H1_0 \ldots H1_n$	$H1_n = /, I(=I, x(\in H1_{n-1}\, x)x)$
$D2_0$	2	E, G,	J = K	$H2_0$	$H2_0 = \hat{}\hat{G}\,\hat{}/i\,\hat{}ij\hat{G}\,\hat{}\hat{}ij\,\hat{}ij$ $[= D2_0\,\hat{}ij]$
$D2_1$	3	E,	H, J, K	$H2_1'$	$H2_1' = \hat{3}✗''/i\,\hat{}ij,\ ✗''/i\,\hat{}ik,\ ✗/\,\hat{}ij$
$D2_1$	3	E		$_\cup H2_0 H2_1 = H2_1$	$H2_1 = H2_n$ with $n = 1$, see below.
$\underline{D2}^*_1$	5		–	$_{\cup\cup} H2_0 C2_1$	$C2_1 = /Ç, d(\leqq H2_0 d)$
$\underline{D2}_2$	10	E		$_\cup H2_1 H2_2$	$\bar{H}2_n = /✗, H(=\text{''}\bar{H}2_{n-1}\,2//, g$
$\underline{D2}_3$	137	E		$_{\cup\cup} H2_1 H2_2 H2_3$	$(= g✗, h(\in Hh), g(\in \bar{H}2_{n-1} g)))$
					where $\bar{H}2_0 = \hat{3}/i\,\hat{}ij\,\hat{}ik$
$\underline{D2}^*_4$	$17_{10}37$			$_{\cup\cup\cup} H2_1 H2_2 H2_3 H2_4$	
$D3_0$	3	E,	H, J, K	$H3_0$	$H3_0 = \hat{3}/i,\ \hat{}ij,\ \hat{3}hij$ $[= D3_0\ \hat{3}hij]$
$D3_0$	3	E,	H, J, K	$H2_1'$	$H2_1' = \hat{\cup}''/i\,\hat{}ij,\ ''/i\,\hat{}ik,\ \hat{}\hat{}ij/\,\hat{}ij$
$\underline{D3}^*_1$	10		–	$_\cup H3_0 C3_1$	$C3_1 = /Ç, d(\leqq H3_0 d)$
$D\infty_0$	$(A : \aleph_0)$		J	$H\infty_0$	$H\infty_0 = /, p(\leqq D\infty_0 S \in \mathbf{Q} S, = \text{‘}S\aleph_0, \in Sp)$
	$(B : \aleph_0)$	E,	J	$H\infty_0$	
	$(C : 2^{\aleph_0})$	E,	J	?	
$D\infty_n$	$(B : \aleph_0)$	E,	J	$_{\cup \cdots \cup} H\infty_0 \ldots H\infty_{n-1} C_n$	$H\infty_n = , p(\in, S(\leqq D\infty_0, S(\in \mathbf{Q}, S(= \text{‘}S\aleph_0)))$
	$(C : 2^{\aleph_{n-1}})$	E,	J	?	$, p(\text{'}\in H\infty_{\,\text{'}i(i<n)} p))$
					$C_n = /Ç, d(\leqq H\infty_{n-1} d)$
D!	$(B : \aleph_0)$		–	$H\infty_\infty$??	
	$(C : > 2^{\aleph_n}$				

PART II

APPLICATION

Having set out in Part I the basic principles of the theory of indistinguishables, and developed the mathematics to the point where the more important and relevant theorems have been proved, I turn in Part II to consider its application. As already indicated, the application I have in mind arises from the unexpected finding that only a relatively restricted class of triparitous assemblages are 'rational', that is, capable of empirical validation. The kind of physical theory which might be built on this foundation has been briefly discussed in Chapter III, where it is made clear that its relationship to empirical data is an unusual one. It seems therefore appropriate to introduce this part by a rather more detailed discussion of this question.

The opening chapter will contain a statement of a Hypothesis, suggested by the theoretical result mentioned above, and an outline of the guiding principles which I shall follow in comparing the hypothesis with actual data. A considerable variety of inferences will be found to follow from it, most of which appear to be correct in the light of present knowledge. But there remain a few undecided points, which may in the end turn out to refute the hypothesis.

CHAPTER VIII

HYPOTHESIS AND PRINCIPLES

A major part of the present chapter will be devoted to working out in useful detail the so-called 'principle of coherence' mentioned as an important aid in assessing the theory in Chapter III. Unless one begins with a grasp of the whole of the hypothesis which is being tested, one will be inadequately briefed on what sort of modifications are legitimate to make, in case the initial results are unfavourable. Moreover, the fact that the subsequent interpretations are not arbitrary, but strongly motivated by the coherence pattern, might not be apparent without preliminary exposition. I begin, however, with a few more general considerations.

8.A. THE ROLE OF THE OBSERVER

The theory has for its primary reference objects which, as I have argued, are inherently unobservable, except indirectly, and as classes rather than as individuals. Nevertheless it is claimed that many points in fundamental physics are what they are because they reflect this Inchoative Plane, which is in itself inaccessible. It is obvious that, if this claim is to be made into a testable hypothesis, some unfamiliar positions will have to be explained.

The claim that the PP is 'constrained' by the IP rests largely on the gaps which the latter reveals: certain numbers do not occur in the PP, that is, are not observed. Without observers, this is meaningless; it is in the pattern emerging from the exploration of their world by the observers that the constraints of the IP appear, and without such observers there is no pattern. One could easily get into a typical metaphysical tangle by pursuing this line. It is therefore expedient to separate, as early as possible, those aspects of the theory which call for explicit recognition of the observer, from those which can be stated, in the proper scientific manner, without reference to him.

This can be achieved as follows: (a) the assumption that the IP 'exists' is meaningless unless the world which the assumption is designed to explain contains observers; (b) well, here we are – so let's assume it; then we find that (c) we can, as it happens, state our basic hypothesis without reference to ourselves, provided (d) that the interpretation of the theory finds a place for the *process* of observation, among the physical phenomena to be explained. Only the last two clauses call for further thought here.

The possibility of stating our hypothesis without observers derives from the argument of 3.D, where the requirement for a rational Sort to have a biparitous homomorphic representation *in terms of its own elements* is argued, not as necessary to convince an observer, but on logical grounds. In particular, failure of the last point would imply that at least some extraneous factors were involved in whatever physical phenomena might be being explained, other than those arising from the IP system in question; and this, we presume, can be avoided by proper care in selecting the phenomena to provide physical correlates. It is not of course out of the question that a phenomenon could express joint constraints from two or more IP systems (i.e. rational Sorts); I have taken no account of this possibility in this work, but it might, with proper caution, help to extend the scope of the theory. For the present, however, it is sufficient to note that, following 3.D, we need not introduce observers into the formulation of our hypothesis.

This leaves us with the final point (d). If no place could be found in the theory where the process of observation could be seen as a legitimate interpretation, it would consign this process to one of the 'gaps' in the PP, that is, forbid it. It could of course be objected that orthodox physical theory is, for an analogous reason, also self-contradictory. But the defence in this case is adequate, if only just adequate: since the aim is to explain the phenomena as existing independently of their being observed, there is no obligation to include these supernatural beings in the explanation. Of course, they have to be allowed in at the end, to test the theory which emerges by suitable experiments; the overall exercise involves an implicit dualism, the more obvious because the experimenters have to be *convinced* – a subjective experience! But at least we know at what point we have to open the door. In the theory I am expounding here there is no such point.

8.B. FORMULATION OF THE HYPOTHESIS

The following hypothesis will serve as a statement of what I propose to test against such empirical data as can be marshalled in evidence for or against it:

H: Physical phenomena will be observed

H1: such that

H1.1: each is governed by a number of degrees of freedom equal to the cardinal of one of the Rational Sorts listed below; and that

H1.2: they fall into classes with different properties, corresponding to the Families of Rational Sorts listed below; and that

H1.3: those corresponding to Perfect Rational Sorts, of which there is one in each Family, will have distinctively simple correlates, as compared with those which realize the Mixed Rational Sorts which predominate in most Families;

H2: where the Rational Sorts referred to above are, with their cardinals:

In Family D0, I Sort: $D0_0$ with 0 members;
In Family D1, $\aleph_0$ Sorts: $D1_0$ with 1 members, $\underline{D1}_1$ with 2, and in general $\underline{D1}_n$ with $n+1$ members ...
In Family D2, 7 Sorts: $D2_0$ with 2, $D2_1$ with 3, $\underline{D2}^*_1$ with 5, $\underline{D2}_2$ & $\underline{D2}^*_2$ both with 10, $\underline{D2}_3$ with 137, and lastly $\underline{D2}^*_4$ with 1701411834604-6963173168730371588410 5864;
In Family D3, 2 Sorts: $D3_0$ with 3, $\underline{D3}^*_1$ with 10;
In Family $D\infty$, $\aleph_0$ Sorts: $D\infty_0$, $\underline{D\infty}_1$, ... $\underline{D\infty}_n$ ... all of which are infinite: and possibly
In Family D!, 1 Sort $D!_0$ with a cardinal not less than that of any other Rational Sort. (8.1)

As promised, there is no mention of observers; the opening phrase refers, however, to observation, which is inevitable, since to predict unobservable phenomena is a mere impertinence. Nothing in the statement of the hypothesis implies that there are no phenomena not reducible to those included. It will be assumed that the two Sorts in Family D3 are entitled, by virtue of this membership, to separate interpretations from their isomorphs in Family D2. Special considerations apply to the two last Families; $D\infty$ contains infinite Sorts, of which for reasons to be explained I shall consider only $D\infty_0$, and in the case of $D!_0$ there is a *real* question of whether it exists at all.

8.C. SOME GENERAL PRINCIPLES

In the remainder of this chapter, I shall set out certain principles and strategies useful in guiding the process of comparing the predictions emerging from the hypothesis with known facts. What I here call 'principles' are really only rather elementary theorems concerning indistinguishables as they relate or fail to relate to empirical data; but since they form no part of any systematic exposition I prefer not to use the term 'theorem'. None of them are at all hard to follow.

> No observation provides any evidence as to whether given entities are indistinguishable (8.2) Pr.1

In effect this is merely a restatement of the conclusion of Section 3.3. It is mainly to be examined for what it does not say. It does not preclude the possibility that entities might be *proved* indistinguishable by some theoretical argument; for that is not observation. It also does not refer to classes of indistinguishables, some of which (i.e. rational Sorts) may be observed, or at least evidenced from observation.

> Given two specified entities, to demonstrate that they are distinct requires information in excess of that involved in their minimal specification (8.3) Pr. 2

It will be observed that Principle 1 leaves open the possibility that there are no indistinguishables at all. Since any valid concept of indistinguishability does not rest, as does that of equivalence, on the mere want of information by which distinctions might be established, we should seek to state that there are (or at least may be) indistinguishables by bringing in information. This Principle 2 does. To deny it is to assert that distinction of entities can always be established from their 'specifications' – which is true in a biparitous context, but rules out the existence of indistinguishables. If, secondly, we suppose that distinction could be proved sometimes but not always, the question has three possible answers: they are identical, they are distinct, and no decision. If the third is identified with twinship, we contradict Principle 1, so the meaning must be no more than a failed instance of the first alternative. But if distinction could never be proved without additional information, we can accept the existence of indistinguishables without offence to principle 1.

It might be thought that principle 2 denies the possibility of things being *specified* as distinct. In a mathematical context one can say 'let B be an entity distinct from A' quite legitimately, because the things referred to are abstract items in a theory. But in an empirical situation this could not be a minimal specification of the entities. If 'this' and 'that' are indistinguishable when thus minimally specified, we must still presume that they have different locations which will allow them, as they stand, to be distinguished. This presumption could be verbalized as 'this one here' and 'that one there', the additional information being of the locations. It is noteworthy that, in the above explanation, all three parity-relations come in together; if this were not so, one would expect that the theory of indistinguishables could be formulated by adding axioms on the third parity-relation to those of biparitous mathematics; but this notion is disposed of in Chapter II.

My next 'principle' is a very frequently used one in the sequel; it is an elaboration of Principle 2, but calls for separate statement:

> To demonstrate that a many-valued quantity has a non-limiting value requires information in excess of the minimum required to specify the mensurand (8.4) Pr. 3

Firstly, a limiting (i.e. extreme) value as against a non-limiting one constitute, in any given case, two entities in the sense of Principle 2; their specification, if it is such as to allow for other values, must indicate at least by implication how the values might be ascertained, and must therefore be more elaborate than one designed merely to distinguish a limiting value from the rest. If obervation does establish the limiting value, this extra will not have been needed, but if it does not, we have still to distinguish between non-limiting values, and this involves a further recourse to Principle 2, thereby exceeding the minimum information needed to specify the mensurand, which need not give more than two values.

We now go on a slightly different tack:

> All information is finite in quantity (8.5) Pr.4

This hardly needs proof: sufficient is to point out that infinite information would take infinite time to handle, precluding any publication of the results. Together with Principle 3, it gives the following pregnant corollary:

> Measured values of any quantity capable of an infinite set of possible values are subject to a finite error (8.6) Pr. 5

The consequences of this notion will be pursued in Chapter XI.

8.D. THE PATTERN OF THE FAMILIES

Turning now to more particular questions of strategy, we have to consider what kind of interpretations can be sought for the different rational Sorts. We have here the guidance indicated in (8.1) by the three clauses under H1; in particular, H1.2 and H1.3. I shall consider first what classes of 'phenomena' can be associated with which families of rational Sorts.

The families exhibit a well-marked palindromic arrangement, being respectively singular, infinite, finite, infinite, and singular; we shall find some hints of complementarity in the interpretations attached to the two wings. But first there is a basic question which gives us a firm starting point: if there were nothing in the table of rational Sorts suggesting an unconditionally observable phenomenon, we would expect that the physical plane would be subject to an irresoluble relativism. Since on the contrary we do have at least the illusion that some things are directly observable, we must look for somewhere where the rigour of the tri-paritous regime is relaxed.

And we find such a place in the family D1. The Rational Sorts of this family are all *ordinable*; $D1_0$ being singular is also perfect, but the rest are unions of successive numbers of distinct singular Sorts, and therefore contain no twin members. These RSs can therefore legitimately support an interpretation in terms of things unconditionally observable. Now there is general consensus that all observations in subatomic physics are in essence observations of interactions between particles, that is of a certain class of 'events'. I shall from now on use the term *event* exclusively in this sense. The correspondence between this infinite family of ordinable RSs and the presumably infinite class of observable events is satisfactorily unique and can be taken as our first determined interpretation. It will be followed up in some detail in Chapter IX.

Besides a category of unconditional observables, we also need, as argued above, some place for their correlative, the process of observation.

This needs further thought, but the first family to look to might well be the other infinite one D∞. This will be examined in the next section. Finally, we hope to find in our theory some basis for the various particular quantities, notably the dimensionless ratios which seem to play a fundamental part in physical theory; these look most likely to emerge from the family D2.

8.E. THE NATURE OF OBSERVATION

I start by assuming that observation is a physical process. This is not obviously a complete description, since there is in practice a seemingly irreducible subjective element in any observation; but to be faithful to the canons of scientific method this must be ignored. It may be that to say that observation is a physical event, not usefully analysable further, is sufficient; if so, we can carry on doing physics until we eventually collide with biology. I call this simple view Option A; it will be our main guideline hereafter.

However, it is obvious that observation does involve a complex *process* of operations, which Option A chooses to regard as irrelevant for our purposes. This could be wrong. We should then have to note that each step in the process of observation is itself a physical process, and therefore observable. This obviously leads us into an infinite regression of formally similar physical events. The presence among our RSs of the family D∞, likewise infinite, suggests that this option is also tenable, and I call it Option B.

This too glosses over a difficulty, that most physical processes do not involve any infinite regression; so that to observe an observation is not just another physical process, but somehow different. And each further step in the regression is likewise somehow different from its precessor. This way of looking at observation leads to an infinite sequence in which each term is more complex than its precessor. I shall call this last Option C.

Referring back now to Section 6. K, we recall that in discussing the possible representation of infinite RSs, we could adopt three different treatments, which I there also labelled A, B, C. Treatment A, arising from a strictly 'Intuitionist' standpoint, reduced the family D∞ to the one Sort D∞_0, thus matching our Option A above. Treatment B, applying Intuitionist scruples to representations but not to the triparitous systems represented, gave us an infinite family D∞, in which all the RSs had a cardinality of $\aleph_0$; this clearly matches Option B. Finally, abandoning altogether artificial restrictions on the use of transfinite cardinals, we came up with Treatment C which gave D∞ as a sequence of RSs each of which included the power-Sort of its precessor, thereby providing a satisfactory model for Option C.

This congruence between the ways of considering observation and the possibilities of representing the RSs in D∞ strongly reinforces the initial suggestion that the family D∞ is concerned with the process of observation. It remains to consider how much of this needs to be taken into account in relation to the physical plane.

It is stated in the item H1.1 of (8.1) that every RS ought to find some physical correlate. If we accept either Option B or Option C, the interpretation of the non-initial Sorts of D∞ must be found in some step of a recursive series of 'observations'. Now if we take this seriously, it seems that this way of looking at observation exhibits the involvement in it of consciousness, which has the same recursive structure. I therefore take the view that these options necessarily trespass on territory which is properly speaking outside the competence of scientific method; on these grounds, I propose to consider only Option A in the sequel. But it is not permissible to assert that the other options are invalid; the most acceptable mathematical position is that of Treatment B, though many would not cavil at Treatment C either, and either of these present D∞ as an infinite family. This means that, if the Hypothesis of (8.1) is correct as stated, some aspects at least of 'consciousness' must be counted as physical realities. That may provide food for thought for philosophers. At this point, I merely note the fact, and give notice that I shall not pursue it further.

8.F. UNCONDITIONAL OBSERVABLES

We have so far said nothing about the families D0, D2, D3, and $D^!$. By elimination, these presumably refer not to things which are unconditonally observable like D1, nor to observation itself, like D∞, but to conditional observables.

For most purposes, D2 and D3 can be taken together. They are distinguished from the rest in being finite and non-singular, and in their RSs having rather arbitrary-looking cardinalities. In this respect they seem well suited to lead to the particular numerical constants so prominent in physical theory; and one at least, $^{c}\underline{D2}_{3}$, needs no introduction. The detailed treatment of these RSs is left to Chapter XII, and will prove to be one of the most fruitful parts of the theory. That they all represent things which are observable only indirectly as classes of indistinguishables will there appear plainly.

More metaphysical are the two outlying families, D0 and D! Both these are singular, with one RS each. The latter, introduced to validate the use of unions in constructing other RSs, is fairly obviously correlated with the totality of the physical world. If anyone questions the meaningfulness of calling this 'observable' I am quite happy to agree. It is doubtful whether $D!_0$ can be regarded as strictly a rational Sort; it can have no representation unless we adopt the Treatment B of Section 6. K for $D\infty$, and its legitimacy is thus marginal. One might argue that this too is not unreasonable. Treatment A in effect confines the theory to the physical plane, and consistently with this ignores the Whole, which is obviously more extensive, including for instance the organic plane &c. Treatment C, on the other hand, identifies observation with consciousness itself, and at the same time pushes $D!_0$ beyond the limits of mathematical consistency – into an appropriately theological horizon!

$D0_0$ encounters none of these subtleties; we have no choice within our Hypothesis but to accept it is an RS requiring correlation. The complement of the Whole is presumably Nothing. The implication is that the physical plane is not experienced as a Parmenidean plenum. One can go no further than this without considering what it actually is experienced as, which I shall do in Chapter IX. Here I would only note that the current speculation that the whole of space may be filled with virtual particle-pairs comes pretty close to Parmenides' picture, and is to that extent contradicted by this interpretation. There must be emptiness.

8.G. PERFECT AND MIXED RSS

In principle there would seem to be an important theoretical difference between the initial Sorts of each family, which (except $D\infty_0$) are perfect Sorts and contain no pairs of distinct elements, and the subsequent mixed Sorts, some of whose members are distinct and others twins; the former only are exclusively classes of indistinguishables. This difference we would expect to see reflected in the character of the interpretations accorded to each.

We cannot predict, however that the composite structure of the mixed RSs *must* have an observed consequence. It is, within the theory, secondary to the recognition of the whole Sort as rational, and in almost every case the perfect Subsorts of the mixed RSs are not themselves RSs; nevertheless, it is clearly worth looking out for secondary distinctions within the compass of any correlate we may entertain.

Now there are certain classes of phenomena which require for their description sets of descriptors which are completely symmetrical, and form ordinable or labelled sets only when correlated with specified physical measurements. Prior to such correlation, such descriptors are indistinguishable, and constitute a perfect Sort. It therefore appears that they should correspond with the perfect RSs, of which $D1_0$, $D2_0$, $D3_0$, and $D\infty_0$ are non-empty and this in turn suggests that the descriptors in question are primarily spatial coordinates though with a large query over $D\infty_0$, which will be taken up in Chapter X.

If this is correct, analogy would further suggest that if any mixed Sort should be correlated with a set of descriptors these should not have the symmetry and homogeneity possessed by spatial coordinates. Indeed, we may expect that in some of these cases a correlate involving objects, of which the appropriate numbers constitute an assemblage with critical properties, is more natural than one in terms of components of a description. These cases arise in fact only in families D2 and D3, where we have already found cause to look for the origin of particular physical constants. Correlations of this kind will in due course emerge; and in some cases, the expected classification of descriptions corresponding to the perfect Subsorts of the RSs plainly appears, reinforcing the correctness of the interpretation. It is perhaps not unexpected that no such classification is as yet apparent where our correlate is a class of objects.

Looking back over this brief survey of the 'pattern of coherence' which I have made part of my initial Hypothesis, it becomes clear how tight a fit it requires between the theory and any acceptable interpretation. It is in fact quite remarkable that any such interpretation should be available, and in consequence one can easily be misled into overweighting its evidential value. In fact, I see no way in which this factor of 'coherence' can be statistically evaluated; while one can hardly doubt that it should carry some weight, we must rely in assessing the correctness of the theory on its quantitative 'predictions', relatively few though they be. In my opinion these are in fact sufficiently striking to justify publication of the theory; but the unfamiliarity of the mathematical treatment and the novelty of the philosophical underlay remain powerful obstacles to its ready acceptance.

CHAPTER IX

EVENTS IN THE VOID

In this and the following chapters I shall examine the two classes of rational Sorts (which I shall from now on generally abbreviate as RSs) which seem to lend themselves most readily to correlation. These are the family D1, which I have argued must be correlated with entities directly observable, and the initial Sorts of each family, which in Section 8.G I suggested should have some relation to spatial coordinates. Together, they lead us to a general picture of what we may call the 'shape of the universe' and some indication of the 'nature of its contents'. But the present chapter will take us only half way to such a description.

9.A. THE VOID

I begin with $D0_0$, the empty RS. There is naturally not much to be said about this, and most of it has already been said in Chapter VIII. Its correlate, a realization of nothingness, cannot be identified with space: we have as yet no concept of space, and it is misleading to think of this as a simple or primary concept. However difficult we may find it to visualize a Void without supposing it to be extended in some kind of space, that is what we must do until we have developed the essential conceptual ingredients of 'space'.

The only function of the Void as such is to separate things. We have, in the family D1, a satisfactory basis for assuming an infinite discrete ordinable Sort (that is to say an infinite Set) of entities. Their discreteness does not of itself imply separation; but this is now supplied by the Void with which we correlate $D0_0$. We can thus speak of entities being 'in' the Void, but only if we forget the topological as well as the geometric nuances which ordinarily attach to this preposition. I may say that I am using the word 'in' a highly abstract sense.

Besides this basic use of the Void, its presence in our scheme does have a few logical or philosophical consequences. It is for instance presupposed by any discussion of the expansion of the universe, which I will bring in in Chapter X; it follows that its absence can be held to be refuted by the observations that establish (as most agree) that this expansion is real. But it is *not* space, let alone 'empty' space.

9.B. EVENTS

All physical observations, at least on the smallest scale and arguably on every scale, are of interactions between particles. Not that we have any right to mention 'particles' at this point, for these are not themselves directly observed, except when undergoing interactions; they are no more than the hypothetical subject of the verb 'interact'. Therefore I call what we observe 'events', and in this chapter I can go no further. The case here is different from all the others we shall deal with later, for the family D1 contains only ordinable RSs, which can be individually specified by their ordinals and not merely, as in other families, by their cardinals. Each of these RSs is correlated with an 'event-slot', into which an event, assumed to be in principle observable, can be fitted. If the events could all be actually observed, they would of course be in themselves an ordinable Sort; but in fact of course we can never observe more than a minute fraction of them, and so we can't, as yet, assume that none of them are indistinguishable.

We must therefore start our discussion of the events on the assumption that, prior to their assignment to particular slots correlated with some particular $D1_n$, they may be either distinct or twins. We may assume that they are equally numerous with D1, that is, that there are $\aleph_0$ of them. But for all that, it seems likely that very few pairs of events will be twins. A given event is characterized by a large number of descriptors, some of which define the 'particles' assumed to participate in the event, but others are conventionally assigned values out of an infinite range of possibilities. Thus, the total energy of an event is supposed to be a continuous variable, as are many other of the measures commonly used to describe events. But at this stage of the argument we have no way of discussing such things, and in fact all the descriptors to which it is in any sense plausible to assign a continuum of values turn out to be coordinates in some kind of 'space'. Before we have constructed such a space, we cannot appeal to the position of an event in it as a distinguishing feature of it; and it thus become more reasonable to expect that in themselves (i.e. without assignment to a particular $D1_n$) events may sometimes turn out to be indistinguishable.

It is therefore to be expected that among the (presumably) $2^{\aleph_0}$ *orderings* of the events, some will be twins. A major source of such twinning is evident from relativity theory. Two events so placed that they cannot both lie on the path of one photon cannot both be observed simultaneously;

I shall prove below that two orderings of the events which differ only in the respective positions of a pair of 'mutually elsewhere' events are twins. However, when the events have been all assigned to some particular $D1_n$ they will all be distinguished by such assignment. I shall call the class of events so treated U, and we come now to our first conclusion regarding our RSs:

> The totality U is a denumerably infinite class of events separated by the Void, having an initial member and capable of simple ordering (9.1) C

If then U is an infinite class of events, it can equally be represented as an infinite sequence of finite subclasses. Such a presentation has the advantage of enabling us to consider the orderings of U in terms of those of its segments S, which being finite are more tractable than the infinite totality. That the segments must all be finite is easily proved. For suppose that one of them S_n were infinite. Then the ordinal of the subclass of U including all the segments through to S_n is not less than ω, and (if n is finite) the ordinal of U as a whole will be not less than 2ω. But by construction U has the same ordinal as D1, namely ω; therefore no S_n can be infinite, Q.E.D. Therefore

> U can be represented as a denumerably infinite sequence of finite segments S (9.2) Th.

Note that it is an inescapable consequence of the Hypothesis (8.1) that U is infinite, since D1 is so. But this is not a 'realized' infinity, as will appear in C.2 below, so that there should be no philosophical objections to this conclusion.

9.C. ORDERINGS OF A SEGMENT OF U

We have seen that the orderings of U may be a triparitous class. Since that of the segments S of U is clearly ordinable, only the orderings of any one segment may be triparitous. We now have to determine the conditions under which two orderings of a segment S can be twins. I shall prove that

> If there is a decidable parity-relation between two orderings of a segment S, differing only in the relative ordering of two events whose parity is unknown, these permutations are twins (9.3) Th.

Let the two events concerned be a, b; and let the relations P, Q be $P \Leftrightarrow \overset{!}{-}\,''ab''ba$, $Q \Leftrightarrow =''ab''ba$. As the parity-relation between the permutations is assumed to be decidable, both P and Q must be decidable. Now (a) the conjunction P, Q is self-contradictory, so that $=(/)P,Q$; (b) $'P,Q \Rightarrow Q \Rightarrow =ab$, so that the two events are one, *contra hypothesi*; therefore $=(/)'P,Q$; moreover (c) $P,'Q \Rightarrow P \Rightarrow \overset{!}{-}ab$ (from Theorem (4.31) in Part I), but since this is assumed to be unknown, we have $\nmid\ ()\ P,'Q$. But if both P and Q are decidable, this reduces to $=(/)P,'Q$, so that (a)(b)(c) are all false. Therefore, we must have $=()'P,'Q$ which $\Rightarrow \overset{\wedge}{-}\,''ab''ba$, Q.E.D.

I mentioned above that a major source of examples of this theorem derives from cases where a, b are mutually elsewhere; that is, where no time relation can be accorded to them. This gives us a hint that the ordering of U may have some connection with time. But in this case the orderings themselves are equally indistinguishable, and no consequences follow. But it is obviously possible, indeed inevitable, that we may be unable to establish the parity-relation between some events which do have a definable time-relation; since observation of an event is itself a series of events, almost all pairs of events will be in this class. So that the situation of Theorem (9.3) can occur, in principle, even when the permutations involved are in some sense observable (in reality, only for very restricted subsegments).

Here there is an important consequence. For the class of all mutually twin observable permutations of any set of events must be a rational Sort and, since it is also a perfect Sort, it must be one of $D0_0$, $D1_0$, $D2_0$ or $D3_0$. It could not be $D\infty_0$ because an infinite class of permutations of a finite discrete set of events separated by the Void is impossible, therefore unobservable (which does not follow if they are contiguous, for one can in some sense 'observe' an infinite set of points in a finite region of a dense space). We have therefore proved that

> At most three twin orderings of a segment S can be simultaneously entertained as observable (9.4) Th.

9.D. SPECIFICATION OF PARTICULAR EVENTS

The wording of this result is deliberately vague; the meaning of the existence of this maximum of three twin orderings of S must be translated into more operational terms, before we can do better. Now all along I have been assuming that the events in U can be correlated with the RSs in D1, and this implies that a given event can be specified by its corresponding $D1_n$. But this leaves unanswered the questions How is this to be done? By placing a limitation on the possible orderings of S, Theorem (9.4) has a bearing on this question.

Let the events of S be divided into p subsegments $E_1 \ldots E_x \ldots E_p$ such that each event in any E_x is distinct from every event in any E_y where $x \neq y$, the parity-relations between events in any one subsegment being unknown. Theorem (9.4) indicates that such an ordering can be set up in up to three different ways, all mutually twin to each others; which can be effected, according to Theorem (9.3) only by exploiting the unknown parities within each subsegment. So let a second ordering of S into q sub-sub-segments be set up; in respect of a given E_x this will divide it into a sequence $E_{x1} \ldots E_{xy} \ldots E_{xq}$, once again with unknown parity-relations within a given E_{xy}. Note, however, that such a second ordering may not be possible for some E_x if it is to be twin to the first; this would imply that E_x contained no pairs of events of unknown parity, and since it has been constructed so that all its internal parity-relations were unknown, it must have now only one event. If E_x had more than one event, E_{xy} may now have only one, and again we must stop. But Theorem (9.4) allows that we may be able to find yet another twin ordering of S, under which a given E_{xy} is further divided into sub-sub-sub-segments E_{xyz}. Any further orderings *must* now be distinct from these first three, from which it follows that E_{xyz} must be reduced to a single event. We conclude from this argument that

> Any event can be specified within a given segment S of U by its position within each of three twin orderings of S (9.5) Th.

The degenerate cases, if they occur, are manifested as a lack of free choice in one or two of the 'positions' specified.

This represents a considerable advance towards a means of actually being able to specify an event. Previously, we only had the theoretical method of citing some $D1_n$, which would be in general quite unascertainable, if only from the inherent inordinability of mutually elsewhere events. Now, we have a Sort of twin orderings of each segment, which we may be able to interpret as referring to realizable measurements of some kind. Moreover, if we take the segments as being each the past moiety of a light-cone with the observer at its vertex, so that their sequence corresponds to the observer's view of time, we can eliminate the 'elsewhere', and at the same time identify each segment by its ordinal position in the appropriate time sequence. Thus

> Any event can be specified within the whole of U by its ordinal position in each of *four* orderings of U, one relating to the observer's time, and the other three being mutually twin
> (9.6) Th.

9.E. ORDINATORS

This theorem suggests the familiar system of four space-time co-ordinates so strongly, that it is necessary to point out that it is nothing of the sort. All we have are four ordinal numbers; there is no hint given as to how they might be determined, still less of any geometry of whatever 'space' they may occupy, or even of its topology. The class of four ordinals is not therefore as yet a vector, for that implies a vector space; yet it is obviously a step towards defining a vector. I therefore propose to call it an *ordinator*.

We have as yet no 'space', except the discrete point set U. Since any Set (unlike any Sort) can be simply-ordered, to assert that U has the ordinal ω really adds nothing to this statement. Theorem (9.6) adds a lot, however, but only provided we can somehow identify the permutations of S (I shall leave aside the fourth ordinate for the present) with actual orderings of real events. In principle, this could be done by associating each of them with some sequence of the ordinals identifying the event-slots by which we interpret D1. To carry this out would call for some kind of total enumeration of the events in a given S, which is obviously beyond reason. But we are not immediately concerned with its feasibility, only with its potential results.

Suppose then that we have an arbitrary event, say of a kind k_0 (for the events are of many kinds, even though not all are distinct), which I shall call O. Now let A, B, C be three mutually twin permutations of S, constituting a rational Sort. We are required somehow to 'identify' A, B, C. Let now P be an event distinct from O, say of a kind k_1; identifying A, B, C will enable us to assign an ordinal position to P in each of A, B, C. In general, P will not be juxtaposed to O in any of these permutations; so let D be a fourth permutation (according to Theorem (9.1) distinct from A,B,C), in which O, P *are* juxtaposed. It now follows, from Principle (8.3)

(Chapter VIII), that the placing of P, if defined as next after O in D, in A, B, C is an act involving information additional to that used in the relevant specifications. Equally, of course, the same is involved in the placing of P, defined by its positions in A, B, C, next to O in D. This means that the assertion that O, P are juxtaposed in D contains empirical information about P. It asserts, that is to say, a relation of some kind between them, expressed for this purpose by the phrase 'juxtaposed in D'.

Two points emerge from this discussion. First is that the ordinators which we can use to specify events are shown to carry information (which is by no means self-evident, considering how vaguely they have been defined up to now); and second, they do so by virtue of the fact that we are supposedly able to identify the twin orderings of S, and predicate of them particular sequences of the D1. This, by the way, is the only case where one RS is mapped onto another, the target RS being in this case ordinable (see Theorem 11 (5.4.1)). The components of the ordinators are thus similar in their function to those of vectors, in that they form a labelled Set, each attached to one of the predicated orderings A, B, or C (so that for instance $\nmid (x, y, z)\ (y, z, x)$), which can be represented in the vector manner as an ordered Set; and that they carry specific information. But they do not still refer to a vector space, a fact represented by the fact that the components are ordinal numbers, not cardinals or rationals.

9.F. DIMENSIONS

It is perhaps improper to speak of dimensions in connection with the system denoted by these ordinators. Nevertheless, in relation to any given segment S they have three, and for U as a whole four, degrees of freedom. If they did not, at least one of them would be determined by the values assigned to the others, and the entire system would be degenerate, in relation to the possibility opened by Theorem (9.4). It is of course not inconceivable that such might be the case in one or another particular part of U. But if so, that is special information; if it were true everywhere in U, it would imply some restriction additional to Theorem (9.4). Again, we cannot deny this *a priori*, but there is no warrant for it in my theory; moreover, we know that in fact it is not the case.

The fact is that the system U, notwithstanding that it is, in itself, a discrete Set of events, appears to us as if it were immersed in a dense 4-space, the spacetime of relativity theory. If this appearance fails at sufficiently small scales, that is only to be expected, since its observability depends on the presence of events, of which we must observe at least five to satisfy ourselves that we have a 4-space, and at least two to be convinced that we have a space at all. But in larger regions, containing more than one event, we are often able to locate events, and express their locations by vectors referring to specified coordinates, of which we can in general distinguish four.

We have therefore the problem of relating these vectors to the ordinators we have so far constructed. This problem has two parts: one concerns the origin of the intuitive denseness of space (which I defer in the main to Chapter X), and the other part has to do with its dimensionality. The latter can be tackled now. We have to map our ordinators into a vector space; they have four degrees of freedom, and the space therefore must have at least as many. There being no basis for assuming that it has more, it appears to us as a 4-space. Moreover, one of its dimensions ought to be peculiar, just as one of the degrees of freedom of our ordinators is peculiar, being derived from the ordering of the sequence of segments of U. There is no RS of cardinal 4, and from the Hypothesis (8.1) it must follow that there is in nature no truly isotropic 4-space.

This is in full accord with our results so far. From the finiteness of each segment S (Theorem (9.2)) we can now conclude that

> Three of the four ordinal components specifying a given event in U are bounded below and above, while the fourth is bounded below but not above (9.7) Th.

This is sufficient to suggest the peculiarity of the time-like dimension in relativistic 4-space; proof that it is fundamentally distinct must await Chapter X. Meanwhile, the obvious analogy of the ordinators with vectors prompts the following reinterpretation of Theorem (9.3), which I do not put forward as a rigorous consequence or corollary of it, and so label only as a conclusion:

> Time has a beginning but no end, and at any time space has a finite extent (9.8) C.

This conclusion is in principle verifiable – the first such that we have encountered.

9.G. THE INITIAL RSS

In the course of this chapter I have arrived at correlates of two of the initial RSs, namely $D0_0$ and $D3_0$. In Chapter VIII it was mentioned that those Sorts which are both rational and perfect might be expected to have interpretations in terms of completely symmetrical sets of descriptors, such as spatial coordinates. Having now arrived, by a more detailed and convincing argument, at just such a correlate for $D3_0$, one is prompted to ask what we can say about the other initial RSs.

The first point to mention is that, since these Sorts are defined wholly by their cardinals, each is a Subsort of every larger one. Thus $\leqq D3_0\ D2_0$, $\leqq D2_0\ D1_0$. This suggests that to interpret $D2_0$ and $D1_0$ we should look for subspaces of the 3-space correlative to $D3_0$. If we take as our prototype for a segment of U a light-cone, which is a 3-dimensional subspace of the 4-dimensional U containing all (and only) the events in which photons are scattered and then observed by a given observer at a certain moment, we note that this can be visualized as a closed surface contracting to a point; the surface is then a 2-space, $D2_0$, and its radial extent is a 1-space, $D1_0$. The latter, although spatial in this setup, can of course be taken as a measure of time; indeed, one instance of such a radius, of universal relevance, is the Hubble radius, whose value is the basic measure of the age of the universe. As a closed surface, the 2-space correlate of $D2_0$ can also be seen as the space of orientations of directed axes, such as axes of rotation, with respect to a given event in U.

Strictly speaking, of course, all these, being subspaces of the *discrete* space of U, are themselves discrete, whereas in practice they are commonly taken as dense (indeed, in deference to the logic of integration, continuous). Each of them has in fact a dense aspect, whose relation to the discrete aspect is essentially the same in all three cases. I shall here briefly mention what seems to be the way forward, which I discuss in detail in the next chapter.

There is one further initial RS, namely $D\infty_0$. Unlike the others, this is a derived Sort whereas the others are elementary; but this need not affect its interpretation. Although derived, $D\infty_0$ is perfect, as was proved in Theorem (5.8). If we look for a correlate of it by analogy with those for the finite perfect RSs, we come up with the bizarre notion of an infinite-dimensional space, whose 'points' are denoted by co-ordinators, each of whose infinity of components has discrete values. There must be at least two of these values, and I shall assume that there are only two; for if there were more, they could be expressed as a set of binary descriptors, and it would add nothing to the dimensionality to recognise the latter as components. The structure is therefore that of an infinite Boolean lattice in which all points are completely equivalent. This structure however depends on some way of identifying the particular components of the infinite ordinators, which since it demands completion of an infinite task must be ruled impossible; in effect, the interpretation of $D\infty_0$ must be as a *disordinate* space (if it can be called a space at all), in which every point is neighbour to every other.

Only one further point need be made here. The space of events U must be considered to be a subspace of this disordinate space, for $D3_0$ is a Subsort of $D\infty_0$. This means that we have to consider the events of U as immersed in this disordinate assemblage as well as having the orderly dimensional relationships among themselves which we have described. It is out of this paradoxical mixture that we have to construct the space between events, if it is to find a place in our physics at all – we are as far from common-sense in this matter as in any other theory of the smallest things.

CHAPTER X

THE TEXTURE OF SPACE-TIME

In the last chapter I proposed correlations for the families D0 and D1 and for the initial Sorts of other families; the conclusion was that, taken together, these rational Sorts imply the physical existence of a structure which I called U, to which we could attach a few of the properties associated with space-time, notwithstanding its essentially discrete character. In this chapter I shall turn to consider the only infinite rational Sort which has a legitimate bearing on empirical matters $D\infty_0$, and with its help I shall elucidate further the structure of space-time, to the point where its more familiar features appear, together with others less familiar, not to say bizarre.

10.A. THE INTERPRETATION OF $D\infty_0$

We have seen that the family $D\infty$ as a whole may be correlated, in a so far undefined way, with the process of observation. Its initial member $D\infty_0$ stands as it were on the frontier between subject and object. If that is right, we should expect that it has something to do with how we perceive the world. I shall claim that this is in fact the case, though it will take some time to become apparent. If 'perceiving the world' is essentially perceiving U, and if the organ of perception is a physical correlate of $D\infty_0$, the gross disparity of structure between these two should have interesting consequences in what sort of picture comes through to us.

I begin by assuming that there is some common reference between D1 and $D\infty_0$; and since the cardinality of the latter is the greater this implies that each element of U should be somehow associated with some point in a 'space' V which is the physical correlate of $D\infty_0$. Thus, while the points of V, since they are correlated with an infinite perfect Sort, must be fundamentally unordered, a subclass of them may nevertheless be ascribed an ordering derived from U. Thus V appears as a not-quite-wholly disordinate structure; though only in the most general sense a 'space', it yet makes sense to ask topological questions about it.

The first point to be made concerns the Void. Between any two events in U there is a gap; that is to say, we can imagine a point between them, a point where there is no event, for the ordering already established for U makes the concept of betweenness well-defined. Now not only the two events, but the intermediate point which participates in their ordering, may be identified with an element of V. This procedure can be interated, and at no point involves in the ordering more than a vanishing proportion of the elements of V. Thus, while U is a discrete space with a denumerable infinity of points, its assumed relation to V results in our postulating a dense sub-space of the latter, apart from which almost all points of V remain unordered. Moreover, because of its derivation from U, the $(3 + 1)$-dimensional structure of the latter will be reflected in this dense space, which I shall denote by W. Thus, beside the *discrete* space U of events, and the *disordinate* space V, we have a third 'space' to take into consideration whenever we have occasion to consider U and V together, the *intuitive* space W; I call it so because it seems to possess most of the properties normally intuited from our uninstrumented observation of the world. As classes, these three spaces are related chainwise: $U \subset W \subset V$.

We must remember that all are of limited validity. Intuitive space is dense, which should mean that its essential structure remains valid at however small a scale; but from its derivation it is undefined for regions containing less than two events of U. In such circumstames, we have to rely on V rather than on W. Of the three, only U is strictly 'reliable', – but it is inadequate for a complete description of physical phenomena.

10.B. DISORDINATE STATISTICS

We are faced therefore with the problem of what happens at the limit, when W ceases to be 'reliable'. Suppose we have two events E_1, E_2 between which a particle passes unobserved. For all that W offers us a dense space in which a straight line trajectory can be described, we can only guess the way it might go; for W is appropriate for the completed journey, but not for any part of it. We have thus to resort to irreducibly statistical procedures, just as happens in Quantum Mechanics; but in our case, we have to infer the details from the initially unpromising concept of the disordinate space V.

The method will be to use the property of disordinacy to derive a probability distribution for the position of such a particle *referred* to the space W (because the physical relations we shall use are commonly stated in terms of this space). I shall assume that W is a vector n-space; in the problems arising in this book n may be either 3 or 4. I shall consider a variate v with the dimension of length, and a function of v, fv, which (again with a view of the applications anticipated) I shall assume to be of the form v^q where q is a real number.

Disordinacy imposes two conditions. One is complete isotropy and symmetry, so that if v is a vector, say $v = (x, y, z)$, then $fv = f(x, y, z) = f(x^2 + y^2 + z^2)^{\frac{1}{2}} \,.\, P(f)$, where $P(f)$ is a probability distribution function satisfying both the metric aspect of W and the disordinate aspect of V. If the latter is ignored, P(f) would be zero except for the 'correct' value of f, whereas if W is ignored P (f) would be a constant. This is an example of the second condition of disordinacy, that no variable may be assigned any *particular* value (other than 0, ∞, or in some contexts 1); constancy is the only possible non-particular value of a function such as P(f). Note that, to preserve referrability to W, the formal distinction between x, y, z must be maintained unless proved dispensable.

The integral of $f(x, y, z)$ over a region R of 3-space can be expressed as a sum of integrals over any arbitrary set of mutually exclusive and collectively exhaustive sub-regions (such as a grid of rectangular boxes). The order of such boxes must be irrelevant because of the disordinacy; this entails that the complete integral be factorizable in respect of x, y, z, in the following sense. Since we assume that $fv = v^q$, we have that $\int_X fx dx \propto \mathrm{X} \cdot f\mathrm{X}$ (where $\int_X$ stands for $\int_0^X$ if $q \geqq 0$ or $\int_x^\infty$ otherwise); eliminating the particular value X by replacing the variable x by x/X, we have $\int_1 fx dx/\mathrm{X} \propto f\mathrm{X}$ and so also $\int_1 f(x, y, z) dx/\mathrm{X} \propto f(\mathrm{X}, y, z)$. Now the above factorizability of the integral means that

$$\int_1 f(x, y, z)\, dx/\mathrm{X} \propto f(\mathrm{X}, k_2, k_3) f(k_1, y, z) \qquad (10.1.1.)$$

where the k's are constants depending in general on the metrics of the system. However they are necessarily particular values of the variables, and must therefore be eliminated; that is, in our case they must all be 0 or ∞, and I shall write $k_1 = k_2 = k_3 = 0$. For the triple integral we then have

$$\int_1 \int_1 \int_1 f(x, y, z) dx dy dz/\mathrm{XYZ} \propto f(\mathrm{X}, 0, 0) f(0, \mathrm{Y}, 0) f(0, 0, \mathrm{Z}) . \mathrm{P} \qquad (10.1.2)$$

where P is the integral of the P(f) subsumed in $f(x, y, z)$ over the whole region XYZ = R. Equation (10.1.2) can be simplified since, once factorized, we need no longer maintain the distinction between x, y, z; I shall therefore write v for all, and V for X, Y, Z, and R for V^3, and make the P(f) term explicit, giving

$$\int\int\int_1 fv\, \mathrm{P}(fv) \cdot dv^3/\mathrm{R} \propto (f\mathrm{V})^3 \cdot \mathrm{P} \qquad (10.1.3)$$

From this we have to eliminate the particular value V, which is achieved in the indefinite form

$$\int\!\!\int\!\!\int fv\,\mathrm{P}(fv)\cdot dv^3/\mathrm{R} \propto (fv)^3\cdot\mathrm{P} \tag{10.1.4}$$

This specifies a relation between fv and $\mathrm{P}(fv)$, which is one-one if fv is of the form v^q, and this relation, by its derivation, expresses the conditions arising from the disordinacy of the space V (onto which W is assumed to be mapped). If W is taken to be a 4-space, the only effect is to replace the exponent '3' in (10.1.3.ff) by '4'.

Let us, with the same proviso, write v^q for fv, and hv^p for $\mathrm{P}(fv)$. We get

$$\int\!\!\int\!\!\int v^q hv^p\cdot dv^3/\mathrm{R} \propto v^{3q}\cdot\mathrm{P}$$

that is

$$v^{3+p+q} \propto v^{3q} \tag{10.1.5}$$

so that

$$p = 2q - 3 \tag{10.1.6}$$

and

$$\mathrm{P}(v^q) = hv^{2q-3} \text{ in a 3-space W} \tag{10.1.7}$$
$$\textit{or } hv^{3q-4} \text{ in a 4-space W}$$

I shall subsequently refer to (10.1.7) as defining the *disordinate statistics* for 3 or 4-dimensional (sub)spaces of W. Numerically, according to their derivation, they will apply *in* W, but when the lack of two events in a problem compels the assumption that the quantities concerned are defined, only in W *as mapped* onto V.

It will be noted that if q is an integer greater than 1, for instance if we are concerned with the volume of radius v round a given point, $\mathrm{P}(v^3)$ is an increasing function of v, which should mean that the volume is infinite (or at least should be taken as such). But this overlooks the fact that for sufficiently large values of v there will be more than one event within the enclosed volume, so that the assumptions underlying (10.1) will not be valid for such values of v. Presumably we shall expect some kind of 'smooth' transition from the disordinate regime to progressively better defined statistics as the number of relevant events is increased. But we must remember that not *any* event is relevant to any given problem.

We may summarize the result of this section in the statement that

In any region of W containing less than two events of U, we

must use the disordinate statistics defined in (10.1.7) in respect of

All relations involving what would be called 'distance' in W, and

All dynamical quantities dependent on such distances (10.2)

All this depends, of course, on the actual existence of a quantifiable measure of distance, which we have not yet established even in the space W. This assumption, which must depend on a correlate in U, I have yet to justify; this lacuna must now be filled.

10.C. THE ACTION METRIC

Two steps are called for: first, we must define one or more metrics in the space, and second we must define a unit of measurement in relation to one of them. The first step is mathematical, the second physical. Neither U nor V has a metric, the first because it is discrete, the latter because it is disordinate. However, if it is to deserve its name, the intuitive space W must be metricizable; that is, it must be possible to define a metric for it. But this has to be proved; most directly, by constructing a metric and proving that it is indeed such.

If W is to be an acceptable generalization of U above a certain scale, there must be some parameter in W whose value over a region R of W tends to equality, as this increases, with the number of events in that region Q of U corresponding to R. In any other case there would be no common reference between the two spaces over large regions; but since whatever we observe as if in W is already existent in U as an event (for, ultimately, events are the only observables), such a lack of common reference would entail that W was either empty or a mere illusion, which if it were so would divorce empirical reality from its physical basis, which would undermine the logical basis of my hypothesis. Now, as we have seen, the content of U is one of so many events, and nothing else; therefore the required parameter in W must be asymptotically equal to this number (or, if we are to take account of the possibility of events differing in 'weight' in some sense, the sum of their weights) over a large region Q, mapped onto R.

The total content of any region of space is best given as its energy-content, including of course mass. But since U is a 4-space, each region Q of U has a time dimension as well as its extent in space, so that the appropriate measure is rather its action content. We thus want something in R asymptotically equal to the action content of its correspondent Q in U. Let R be a one-dimensional sub-space of W bounded by two points (x), (y) where $(x) = (x_1, x_2, x_3, t)$, a 4-vector reflecting the ordinators locating an event in U but with rational-number components, together with all the events in U which are indistinguishably removed from this line; and finally let $A(u)$ be the action associated with such an event at (u). Then I shall prove that

$$M_a(x, y) = \underset{\in Ru}{\text{Min}} \sum_{u=x}^{u=y} A(u) - A(x) \qquad (10.3.1)\text{Df. } M_a$$

is a metric for any two points (x) (y) in W. A 'metric' is defined as a real number associated with any pair of points p, q in a given space, such that (a) $M(p, p) = 0$; (b) $M(p, q) = M(q, p)$; (c) $M(p, q) + M(q, r) \geqq M(p, r)$. That (a) holds for the metric defined by M_a in (10.3.1) is obvious on setting $x = y$ in the summation limits; (b) is identically true if $A(x) = A(y)$, but if not, it is satisfied asymptotically for large numbers of events; (c) is true because $M_a(p, q) + M_a(q, r)$ whatever it is, is a member of the class of values of which M_a (p,r) is defined as the minimum. Finally, M_a (p,q), being a quantity of action, becomes a real number on being divided by any arbitrary unit of this mensurand. It follows that $M_a(x, y)$ as defined in (10.3.1.) is a metric for the space W, Q.E.D.

It is, however, neither a practicable nor a convenient metric. Applied to any Euclidean space in which the distribution of events is non-uniform, the space as measured by M_a is non-Euclidean, because a path avoiding a concentration of events is defined as shorter than one passing straight through it; but our everyday intuitive concept of space is just of this kind, which M_a thus drastically distorts. Of course, there is no reason to expect W to be Euclidean; but there is an obvious recommendation for any metric which will allow us to report that some regions of space are full of lumps which seem to be disposed in a topologically simple manner, and not as a uniform distribution of material in a knotty space.

Now that we know that W is metricizable, we can look for some natural unit of length, or some class of such, valid at the smallest scale, and try whether we can produce a more practicable metric by building on such a foundation.

10.D. THE DISTINCTION METRIC

Since the space W is constructed, notionally, by interpolating between the events of U, which are in general discrete and separated by the void, an obvious choice for such a unit might be the minimum separation between events, which would enable them to be recognised as distinct. That is to say, we have to consider how such a separation might be defined, so as to give rise to a usable metric. I shall prove that

> If e,f are a pair of events subject to a separation $M_d(e,f)$ in W, such that if they were not separated they would be indistinguishable, then there is a finite distance d such that (10.4) C.

$$\leqq d\,M_d(e,f) \Rightarrow \hat{\_}\,ef$$

$$> d\,M_d(e.f) \Rightarrow \overset{!}{\_}\,ef \qquad (10.4)\ \text{Df.}\ M_d$$

The statements in (10.4) are false if either (a) the value of *d* has to be taken as either indefinitely small or indefinitely large, or (b) there is a value or range of values of *d* for which the parity-relation between the events is indeterminate. The case (a) amounts to saying that events are always twin or always distinct irrespective of their separation. That they are distinguished by distance is obvious from the fact that in principle one can look at one such separated event, or at the other, or stepping back at both together, contrary to the definition of twinship. That they can sometimes be indistinguishable is part of the general presuppositions of the present theory. Therefore falsification by (a) fails. If there were a range of values of *d* for which the parity-relation was indeterminate, this would in principle offer an opportunity for a statistical experiment, from which we could infer a *mean* value for *d* (which would suffice well enough for the purpose of defining a metric in W), unless its range were infinite, which would mean that the relations $\hat{\_}$ and $\overset{!}{\_}$ were empirically indiscriminable, again contrary to my axioms. The only remaining possibility is that there is just one value of *d* where this happens; since W is a dense space, there is zero probability of finding such a value, but if it were found, it would answer to what we were looking for, Q *satis* ED.

We may therefore expect that in every case there will exist a threshold of distinction *d*, in general statistically defined, determining whether two equivalent events will be perceived as distinct or not. The probability of distinction must be monotonically increasing with distance, because it is a probability that part of the Void intervenes. However, C. (10.4) does not solve the practical problem of actually measuring distances: we can't do this by putting electrons end-to-end, which is all that it suggests. There are, however, measures of distance associated with any given mass, such as the Compton radius $h/2mc$, which provide the necessary link.

10.E. THE CHAIN OF MEASUREMENT

The formula $h/2mc$ contains the natural constants h and c, which we shall encounter in 11.B and 10.H respectively. Being dimensional, they can be given the value 1, thereby defining units of 'action' and 'velocity'. We can safely assume that the threshold of distinction is related somehow to $h/2mc$, and we can begin by looking for conclusions which will refute (or fail to refute) the conjecture that the two are equal.

Now the Compton radius is commonly defined in terms of centimetres and grams and seconds, and is known, for some particles, with fair accuracy in these terms. It is thus known, *a posteriori*, to be a usable metric, anchored to the 'distinction metric', and linking that, via a complex of observational techniques, to the man-scale units mentioned. These same units are in the opposite direction linked, via a whole series of overlapping techniques mostly within the field of astronomy, with the 'action metric', useful only at the largest scale.

It is at this point that I ought to bring in the so-called 'Cosmological Principle'. I shall argue that

> If the volume v of a region of a given segment T of W is defined by the action metric M_a, then the energy density over such region tends to a common value for large enough v independently of where the region is (10.5) C.

If this were not the case, we could divide the whole of T (and by implication a large part of W itself) into moieties having different mean energy-densities. This would be, however, from Principle (8.4) a statement containing information. Now whereas information can arise through random assortment processes, this is true only on a limited scale; if it existed on a universal scale, it would point to an outside source, which is of course lacking. We are therefore led inevitably to predict (10.5).

So far as is known, the Cosmological Principle holds good throughout the region open to sufficiently close inspection. It should, however, be pointed out that in the form stated in (10.5), it is irrefutable, since, if there were any deviation from it on observational grounds, we would be able to claim that we could not yet extend our observations to a sufficiently large region. However, the principle could be restated, with the proviso that uniformity should be observed over volumes v small compared to the whole of space. If the latter were infinite the amendment would be nugatory, but if it is finite the amended statement of the principle does, as of now, seem to hold.

We cannot, of course, strengthen (10.5) by replacing T by W, and space by spacetime – this strengthened cosmological principle would hold in a steady-state universe. As we shall shortly find, the present theory does not accord with such a state of affairs.

10.F. THE CONCEPT OF A PARTICLE

I have shown that, of the four components of the ordinators of Chapter IX which identify events in U, one is distinct from the other three, even when abstracted from the particular ordinals which may be predicated for given events. I have also proposed that the points of W can be identified, provided that the local disordinacy of V is respected, by a rational 4-vector, among whose components the same distinction can be made. We now come to consider what observable consequences may be expected to flow from this lack of 4-dimensional isotropy in W.

The first point is that the distinction of the fourth component must obtain in regions with > 1 event always and everywhere; for if not there would be points or regions in W where its topological structure was different from elsewhere, thereby introducing local peculiarities into this space, corresponding to the presence of information. We can not argue that such information may not exist locally, but we can legitimately consider first the structure obtaining in its absence, before asking what factors might disturb this structure. This then being assumed, the distinction between the fourth component (which I shall call 'time') and the rest does not hold only between successive segments T of U, but is also a relation between neighbouring *events* in successive segments. I shall call this relation C; since t is simply ordered in U, and since C is understood to be an observable manifestation of t, then C must be at least a partial-ordering relation between events.

Suppose, however, as a special case, that C were itself a simple ordering. In that case, every event would correspond to a point in U having the relation C to one and only one point in the preceding and following segments of U. This would be tantamount to assuming the existence of continuous, even if not continuously observable, entities; we could call such entities 'particles', brushing aside the metaphysical objection that they are thereby presumed to exist while not being observed. We can hardly dismiss this objection however, when we note that if C were indeed a simple ordering our 'particles' would *never* be observed; for in fact our only attested observables are the events in U, and an event to be observed must have either more than one input 'particle' or more than one output 'particle'. We need not assume that *all* events are of this class; but some of them must be so if anything is to be observed at all. We must therefore assume that each potentially observable event has a C-neighbour in each neighbouring segment, *and* at least one other in one of them.

One might suppose that a weaker condition would suffice. There might be 'events' in which 'particles' are created out of nothing, or in which they mutually annihilate each other, leaving no residue, not even a photon. Such events would, of course, violate the conservation of energy over any region containing just one of them, or containing an imbalance of creative and destructive events, with the well-known result that over such a region the laws involving energy would vary with time. The objection to admitting such events is not any logical impossibility, but that if they happen (as in the case of local departures from normal topology in observed space, of which they are a special case) they should be accompanied by some information: e.g., specifying their rate of occurrence. In the absence of such information, we must ascribe to them either a zero rate, or an infinite (or at least a limitingly large) rate. The latter, given that the conservation of energy does seem to hold with high precision, entails that creation and destruction must everywhere be precisely matched; what then would we actually observe? This is, of course, a well-known speculation. It is not so much refuted as ignored in my theory. I therefore conclude that

> Every observable event E has at least three neighbours of which at least one has a smaller t-coordinate and at least one a larger t-coordinate than that of E (10.6) C.

We can now picture the relation C as being the involvement in two C-related events of a common *particle*. This concept strictly adds nothing to what is already implied by the symbol C, but there is a real gain from hypostatizing particles in this manner. As we have seen in the case of indistinguishables, it is often useful to ascribe attributes proper to objects to entities defined in a quite abstract way. We ought not then to be surprised to find causal relations between events interacting with electromagnetism or the strong interaction, or possessing a measurable mass.

'Real' particles certainly help to account for the undoubted persistence of physical objects through time.

C. (10.6) also invites the notion that one particle can have a 'life' extending over many events, which are thus reduced to mere incidents, in which other particles are perhaps transformed or destroyed. So attractive is this picture that we must constantly remind ourselves that particles as such are unobservable. They have no existence except as relations between neighbouring events, as in C. (10.6). They are exceedingly useful as a linguistic device: nouns derived from verbs, as happens in all known languages. Out walking the other day, I came across a bite. My dog got it from another, and sent it back again. I wonder where it is now?

10.G. THE POLARITY OF TIME

According to C. (9.8), there is a first member of the sequence of events U, and no last member of the sequence of segments S. This imposes a polarity on the time-structure of U, and so of W also. Any attempts to observe the effects of this polarity will be the more successful, the larger the time-span involved. If, for every event, we could find another exactly reversing its effects, there would be no polarity; there must therefore be some events more often observed than their time-reversed complements. We can therefore infer that

> Phenomena involving large numbers of events are in general irreversible (10.7) C.

One the other hand, in respect of a single event, which may be considered as occupying a negligibly short time, we must expect to find no means of determining, from such an event alone, what the direction of time actually is. Indeed, the local disordinacy attributed to V carries the implication that in respect of any *single* event, no direction in 4-space is capable of indentification with that of the coordinate t. That is to say, repeating (10.2.3):

> Any region of V containing at most one event of U is isotropic in all four dimensions (10.8) C.

It will be recognised that this implies that there is no absolute distinction between particles and their anti-particles, and that tachyons (particles travelling faster than light) might occur. However, this conclusion conflicts with Theorem (9.4), which insists that no more than three dimensions can be isotropically related. There must therefore be some impossibility attached to the observations which (10.8) seems to allow. And, indeed, it is not hard to see what it is. To establish that one has observed a tachyon, one must have observed *two* events by which to time it, and one is therefore in a context to which (10.8) does not apply. There may indeed be tachyons, provided you don't look; and antiparticles are only an extreme case of tachyons.

I therefore move on from (10.8) with two remarks. One is that things may have consequences, even if they cannot be traced out by observation; I shall use (10.8) again in this manner in Chapter XII. The other is the principle that what cannot be observed needn't be thought about. Thus, one may draw conclusions from the 4-way isotropy of V, but may *not* postulate that some tachyon traverses such a space.

10.H. THE LIMITING VELOCITY

We have seen that we can define a distance metric which is valid for 3-space. But W is properly a 4-space, and we shall not have adequately characterized W until we have a metric applicable in all four dimensions. Given that the fourth dimension of W corresponds primarily to 'time' (originally emerging as the succession of finite segment of U), this problem is solved if we can define a unit of time.

From the principle of dimensions, a unit of time can be defined if (and only if) we can define a unit of velocity. If v is such a unit, the unit of time can be defined as d/v, where d is the already defined unit of distance. The virtue of using velocity in this way is that a given velocity can be represented geometrically by a line of a certain angle to one representing (by arbitrary definition) a point at rest. The wider the angle, the greater the velocity. The common convention, which is as good as any other at this stage of the argument, is that the velocity represented is proportional to the tangent of the angle; on this basis, we can identify our unit velocity v with that angle whose tangent is 1.

If now such a metric exists, the paths of all points moving through an arbitrary point O with unit velocity describe as their locus a hyperbicone with vertex at O.

In a truly isotropic 4-space (with as yet no metric), we can define any four paths through O as being mutually orthogonal, thereby locally and partially defining a metric. Alternatively, given a metric already defined, we can construct a set of four orthogonals. The symmetry of such a set of paths admits of their being correlated with a perfect Sort. But as there is no such sort, though there are perfect Sorts with 3 members, we must find that one of our four orthogonals is distinct from the rest. It must therefore be possible in some way to *label* such a path by means which will be observationally recognisable. Since at this point in the argument we have not yet finally constructed the relevant metric, such a labelling procedure has to be definable in strictly topological terms. If we enclose O in a simple closed hypersurface, each of the four paths through O will intersect this surface in two points with O between them; if then both of the points belonging to one path are enclosed in small closed surfaces, this pair of surfaces will constitute a 'label' for this path such as is required. Lastly, if we imagine the closed hypersurface to be shrunk down to coincidence with O, the two closed surfaces serving us as a label will describe a locus topologically isomorphic with a hyperbicone with apex at O.

Thus, in an otherwise apparently isomorphic 4-space, we should, according to the theory, always be able to attach a physically detectable label to one of any set of four orthogonals through a point, and one way of doing this is the construct a hyperbicone which shall contain one of the orthogonals and exclude the other three (containment of a path *in* a hyperbicone is defined as is the containment of a point within a closed surface). As applied to the 4-space W, this construction can be used to mark the axis of 'time' as distinct from the three 'spatial' axes.

It is, however, a necessary part of the notion of a label that its presence should be, or at least correspond to, something observable. What then is the observational significance of our hyperbicone? In comparing our hyperbicone label with the geometrical hyperbicone described before, it is clear that one answer might be that it separates a class of 'slow' paths from another class of 'quick' paths. If it is to be an observable distinction, however, it is further required that different conditions (a merely numerical boundary presupposes an existing metric, and thus is not acceptable here) should govern the observability of these two classes of paths. At its simplest, one class should be observable and the other not, at least under stated conditions. In principle, such a distinction could operate either way round; but it is obvious that, since objects can move with arbitrarily small velocity relative to each other the real world (as thus experienced) can only be described by giving the hyperbicone the role of an upper limit on observable velocity.

We thus find that one and the same construction serves two purposes namely to provide a definition for a unit of time and to label, in an observable manner, one of the directions in W as the axis of time. The unit is required because 'time' lies outside the symmetry of the three spatial dimensions, and is provided in the process of labelling this 'time' (which must in such a case be possible). In its capacity as a metricizing device, the label *defines* the limiting velocity, which it represents as a universal constant. This constancy could be formally denied, but only at the cost of requiring another less obvious metric by which to measure 'time'. To suppose that the measure of time varies with time is a logically dubious option – but might it not vary from place to place?

It can and does. According to our Principle 3 any particle will travel with the limiting velocity c unless there is some reason why not. Such reasons can be intrinsic or extrinsic. The former include such properties of the particle as inertia – these of course tell us nothing about the limiting velocity itself. Extrinsic reasons for slow movement are on the macro scale such things as friction or viscosity, and on the micro scale, which is what is relevant here, refractive index. Even a photon, travelling through glass, encounters a locally reduced value of the limiting velocity. We therefore define the limiting velocity only in empty space, and expect it to be exhibited even there only by particles of zero inertia. Such are photons. Therefore we identify the limiting velocity as that of light (in vacuo), namely c.

We have thus found means of defining (not of course of determining the magnitudes of) three 'natural' units of dimensional quantities. These are the quantum of action h, the Compton radius of the proton d, and now the limiting velocity c (of these d has many rivals pertaining to other particles, but since the proton is the heaviest simple stable particle, having the smallest 'radius', it has an obvious claim to priority in this respect). In 'natural' units each of these contains '1' unit of the corresponding quantity.

10.J. THE BIG BANG

The preceding considerations lead us to the highly counterintuitive conclusion, which is nevertheless a direct corollary of Principle 3 and Theorem (10.9), that:

> The limiting velocity c is independent of any motion of the observer (10.9) Th.

This provides the starting point (or at least one possible starting point) for the theory of relativity. I shall not follow this up here, since even if it were within my competence there is no current challenge to the basic correctness of this theory as a description of the large-scale geometry of the universe. There is, however, a particular application of it, which has many implications relevant to the empirical testing of my theory. This is that

> The greatest distance observable in W when a time t has elapsed since the first event of U is not greater than $ct/2$
> (10.10) Th.

This is obvious, in the sense that, if it were not true, something must have been observed to arrive with a velocity greater than c, contrary to (10.9). Moreover, according to Principle (8.4) something must have in fact travelled at this velocity (unless there is some reason why not).

It might be thought that (10.10) implies the existence of a well-defined axis of time. This is not so; for if there were, the particles in W would have a well-defined inclination to this axis, and therefore an absolute velocity, which is nonsense. The direction of time *is* therefore always relative to the motion of the observer; but this relativity does not imply any 4-dimensional isotropy of W. For any one observer there is a future, containing all events in a hypercone centred on his own location, and a past containing all events in the opposite hypercone; and of course there are events excluded from both hypercones. These are necessarily unobservable from the location in question, and the existence of this zone of unobservability guarantees the distinction between time and space for each observer, as stated in (9.8). No observer is privileged over any other in this respect, nor does the theory allow for the existence of any universal observer, or God, within the space W.

As mentioned, Theorem (10.10) does not rule out that the actual velocity of expansion of space may be less than c. General relativity theory indeed predicts that there will be some deceleration due to the gravitational field of the contents of W. It is usually assumed that space is topologically the hypersurface of a hypersphere, or if there is insufficient matter to recollapse the world, of a parahyperboloid. But, on reflection, there is little theoretical basis for this assumption, and so far very little empirical evidence; does the present theory throw any light here?

10.K. THE CONNECTIVITY OF SPACE

The first problem is that all of the above pictures seem to conflict with our principles. A closed hyperspherical space and an open parahyperboloidal one both entail specific numerical values for the deceleration parameter $q_0 \neq 1/2$, notwithstanding the want of any plausible source of the information which Principle (8.4) would associate with this inequality. But a flat space if it is the middle term between these two, while it calls for no such 'information', is infinite, contradicting Theorem (9.8).

Let T be a time-slice of W; we need not ask what sort of slice since our enquiry will be topological not geometrical, that is, to consider what sort of axis-system we might define for T. By an 'axis system', I mean any set of n 'families' of $(n - 1)$-manifolds, each family being a simply-connected set of manifolds disjunct from every other family, and n being the dimensionality of the space. It is assumed that every point of the space belongs to at least one and at most a finite number of manifolds of each family. Thus, to within a finite ambiguity, each point is identified as the intersection of manifolds, one from each family. For any vector space such an axis system can always be found; for our T, as we have seen, $n = 3$.

Certain restrictions on the choice of an axis system for T follow from points already established, mainly to avoid contradicting the cosmological principle (not merely (10.5), but wider than this statement of it). These are:

1: Nothing may be dependent on the point chosen as origin of the axis system (10.11.1)

2: No region of T may be topologically distinct from any other (10.11.2)

3: No point in T may exhibit singularity with respect to its neighbours (10.11.3)

4: No numerical parameters are involved (other than those such as n inherent in the problem itself) (10.11.4)

By an 'axis', in relation to such an axis-system, I mean any path through the origin having one and only one point in each manifold of a given family. Since T is finite, it follows from (10.11.2) that all the axes of our system are closed loops; for if there were a boundary involved, this would constitute just such a topologically distinct region, while if not, an open axis would correspond to an infinite family of manifolds and so to an infinite extent of T. Concerning the mutual relations of the axes, we are thus left with only four options to choose between:

1 : All three axes intersect at one point other than the origin (10.12.1)
2 : Two of the axes intersect again, but not the 3rd (10.12.2)
3 : Each pair of axes intersects at one other point (10.12.3)
4 : The axes intersect only at the origin (10.12.4)

The first case is that of a hyperspheroidal space. The second defines a torospheroidal space; the disparity in the topology of the axes makes this option unacceptable. The third case (if it is indeed conceivable) involves 'numerical magnitudes' in the ratios between the distances of the pairwise intersections from each other and from the origin, and is excluded by (10.11.4). The fourth gives us hypertoroidal connectivity. The choice now lies between (1) and (4).

Two properties at least can be found, in which the hyperspherical and hypertoroidal spaces differ. The first concerns the choice of axes. In a hyperspherical space, all choices of axes are, in the absence of a metric, topologically equivalent and so equally eligible; but there is in the present context a metric, defined by (10.5); for if the mean energy density is uniform in relation to one set of axes, it will not be so in relation to a second set differing in mutual obliquity from the first. There is therefore an empirical criterion of orthogonality for such a space, but no inherent property for it to refer to, so that the cosmological principle seems to carry information. In a hypertoroidal space however the number of numerically distinct vectors which denote a given point depends on the choice of axes; an arbitrary choice gives an infinite number, certain choices produce a finite number, and a few (which can be shown to be all metrically equivalent) make the vector specification of each point unique. There is therefore an intrinsic property of the space, to which empirical data on the distribution of energy-density can refer, relieving the cosmological principle of the imputation of carrying information.

The second differentiating property is that in a hypertoroidal space any set of 2-manifolds, one from each family, intersect in one point only, whereas in a hyperspherical space they intersect in two points, thereby providing two map-images of each vector; this is the converse of the multiple vectors for each point possible in a hypertoroidal space. The whole of the hyperspherical space thus falls into two halves, with the origin at the pole of either half, each being a mirror-image of the other. Since the origin is arbitrary (10.11.1), every region of the space large enough for (10.5) to be applicable to it must be devoid of intrinsically chiral configurations. Moreover, if there were regions of *any* size, over which asymmetrical chiralities were observed in respect of events at the smallest scale, we would infer the presence of locally determining factors over regions. This is not a priori impossible; but since they would be manifest at the microscopic level and yet prevail over macroscopic regions, such factors would be comparable to (if not indeed manifestations of) an assortment of the material of the universe into matter and anti-matter. This is generally regarded as highly unlikely; but if there is no such assortment we would expect strict parity-conservation to obtain for all interactions in a hyperspheroidal space, which is known not to be the case. None of this however holds for a hypertoroidal space, where each vector has a unique corresponding point, and the space as a whole has no necessary self-symmetry.

There seems, from both these points, to be evidence in favour of hypertoroidal connectivity (there is more to come in Chapter XII). There is, however, an argument which could be brought against it, that in such a space there is a unique system of axes of minimal length, which we might expect to observe evidence of if we could see far enough. But when we examine the geometry of the system, and ask what we might see which would identify the axes, the answer is Nothing. The geometrical symmetry of a hypertoroidal space is still hyperspherical (a point easy to overlook, if we visualize the space the easy way as a crystal-lattice !), and there could be nothing to mark where the minimal-length axes were. This objection therefore fails.

These considerations therefore suggest, but do not strictly prove, that:

> The connectivity of every segment T of W is that of the hypersurface of a hypertorus (10.13) C.

> No region within any such segment is necessarily symmetrical in respect of chirality (10.13a) C.

CHAPTER XI

THE CONSTITUTION OF MATTER

We have seen that consideration of the RSs of the family D1, along with D0, gives rise to at least a rudimentary concept of space; and that when this is joined with the initial RSs of each family, we arrive at a reasonably full characterization of spacetime as it is known both to physicists on the one hand, and in common experience on the other. In comparison with this wide prospect, I shall in this chapter deal with the two finite families D2 and D3, from which we shall derive a quite different set of interpretations – essentially, the more basic elementary particles and their descriptors (quantum numbers).

I have already mentioned that it may be legitimate to consider more than one 'interpretation' as valid, under appropriate circumstances, for a given RS. We shall here encounter several examples of how this idea works out. The underlying logic is that, when we encounter some situation where we may suspect that some Sort of indistinguishables may be involved, the first question to ask is always 'Is there an RS of the cardinality required by the evidence?' If there is not, then indistinguishables are out: but the theory then requires that there be sufficient evidence, either to support this conclusion, or alternatively to suggest that the cardinal has been wrongly identified. It would be easy to escape notice by not mentioning counter-examples of this kind; it is the more unfortunate that I have been unable to find any, apart from the rather feeble case of the non-isotropy of relativistic spacetime. It is possible however that the coupling constant of the strong interaction may be another. Anyway, there is no reason why the same cardinal should not turn up in more than one manner, as a genuine instance of the effect of the inchoative plane on the physical plane. But the principle of coherence will then lead us to expect some evident connection between the interpretations offered in such cases; and again, if none is found, we must be suspicious of one or both the arguments involved.

11.A. THE FINITENESS OF INFORMATION

I begin by referring back to my Principle (8.5), that information is always finite. This has different effects, according as we are dealing with measurements whose values are may be a dense set, or with discrete quantities. We have met one source of dense values in the statistics pertaining to 'disordinate space'; in this case, the statistical procedure itself takes sufficient account of the finiteness of information. Another class of seemingly dense measurements in those which are defined as ratios. Often these are not strictly dense, but they are usually treated by orthodox statistical methods nonetheless.

Of more interest for our purpose is the discrete case; mainly this is because the observation of events is logically the fundamental operation underlying all observations, including measurements. Events are in themselves discrete, but can also be said to differ in magnitude. The discreteness, which is accompanied by separation by the Void, is in a 4-space, and these magnitudes are therefore more properly measured as action rather than as energy. Information being finite, the magnitudes will inevitably be percieved as discrete, unless subject to statistical dispersion. But we have not in this case either a disordinate or a dense system of values which can be justified a priori because we are dealing with what we take to be irreducible observations. The only justification for a statistical treatment is therefore the suspicion that the apparent size of the 'quanta' might depend on the amount of available information. The character of such dependence might be studied, recursively if need be; but if it is not to reduce to the former case, the recursion must terminate finitely, and so yield a finite quantum, not dependent on the circumstances of any particular observation, underlying measurements of 'event magnitude'.

It is hardly possible to prove a priori that this must be so. A non-quantized world, however repugnant to intuition, is not demonstrably self-contradictory. But the empirical evidence that it is not so is by now sufficiently convincing to permit us to identify the quantum mentioned above with Planck's constant h, which has the appropriate dimensions. We have therefore two kinds of variables open to measurement: those which are quantized and those which are not. Quantized mensurands are either those which have the dimensions of action, or which enumerate elementary quantities (of which later sections treat in further detail); dense mensurands are either such as can be extrapolated into disordinate space V (spatial and temporal intervals), or which are ratios of possibly incommensurable quotients.

11.B. QUANTIZATION AND CONSERVATION

The concept of quantization is closely associated with that of conservation. Those quantities which are quantized by enumeration of objects persisting in time are in general conserved; the reason for this is apt to seem self-evident, until it is realized that some such quantities are in fact not conserved in all types of event. There are, however, principles of a quite different character, which I shall now consider, such as the conservation of energy, which do not depend on enumeration.

It is a well-known result of general dynamics that a system in which dynamical laws do not vary with time must exhibit a conservation of energy; and that one where the laws are uniform over all locations has a conservation of momentum; and that if further the laws of the system are invariant under rotations, there will be conservation of angular momentum. In each case, the product of the dimensions of the transformation and the corresponding conserved quantity are those of action: this in fact provides an alternative argument for the fundamental nature of action measurements, though one which presupposes the separate existence of all the quantities concerned.

If we now consider how, as argued in the last section, measurements of action are quantized, it follows at once that, whenever we measure both of any such complementary pair of quantities on one system at one time, the precision of the two measurements will be inversely related, since their residual errors cannot multiply to less than one quantum. We have therefore a qualitative argument for the well-known Heisenberg Uncertainty Principle (though we are still far from an adequate mathematical account of it).

One might suppose that the rigour of this principle could be circumvented by repeating the measurements concerned many times, and calculating mean values which would have a reduced error. In most of the contexts where the principle is invoked, however, the systems involved are not capable of being observed repeatedly. Exceptions to this are however thinkable: thus, repeated measurements on the angular momentum of a spinning body could be in principle be paired with observations of its axial orientation, and since angular momentum is conserved in a closed system, it could be arranged so that the accuracy of that measurement could be indefinitely increased. If the complementary orientation were also conserved, the same could apply to that. In fact, of course, this is not so; there is no conservation of orientation, nor of any of the other quantities involved in the uncertainty principle as complements of those which are conserved.

Since I shall refer back to them in the sequel, it is convenient at this point to state more formally the conclusions of this section:

Whenever simultaneous measurements are made of a co-referential pair of mensurands whose dimensions multiply to action, the product of their respective errors is not less than Planck's constant *h* (11.1) C.

In any dynamically closed system, those quantities whose action-complements are measured in a system interpreting one of the finite initial RSs are conserved (11.2) C.

11.C. SOME PROBLEMS OF CORRELATION

The reference to the initial RSs in C.11.2, which may be found obscure, has already been elucidated in Section 9.F. These RSs have been correlated with degrees of freedom, associated with simple dimensional quantities. These are time (1 d.f., for $D1_0$), orientation (2 d.f., for $D2_0$), and location (3 d.f., for $D3_0$). There are no analogous correlates for the non-initial Sorts. But C.11.2 suggests that their complementary conserved quantities should be at least associated with the same families of RSs. This association is made explicit in the table in Chapter XIII. It is not indeed very enlightening; but it might perhaps stimulate useful insights in lateral thinkers.

The initial RSs are also, however, open to correlations which originate in the non-initial Sorts, and call for analogical retropolation to the initials. These are of two kinds, which also have been already mentioned, namely descriptors, or components of descriptors, and aggregates exhibiting some identifiable property. I shall consider component correlations in detail in Section 11.6; here I shall deal with the aggregates.

I have insisted in several places that the only ultimate observables known to us are events. Nevertheless we certainly can get what seems to be fairly consistent information about other things which we hypothesize partly to explain the events we do observe. Inasmuch as these indirectly observed entities can be extensively treated by biparitous mathematics, they are also 'observable' as far as the present theory is concerned: that is, they cannot be primary indistinguishables, though they may be equivalent to each other, even inscrutably so (to use the terminology of Chapter III). They could, however, be explained within the present theory as aggregates of indistinguishables, if there were sufficient evidence to make this hypothesis plausible. Such aggregates, which are observed as unitary or elementary particles, I call primary aggregates.

If two such particles of the same kind are composed of two disjunct but twin aggregates of indistinguishables, they will be empirically distinguished by their states only, that is by their different locations, paths, momenta, &c. If in a particular case, these distinctions of state are removed or precluded, the particles will become themselves indistinguishable in the strict sense, and can then be observed only as classes of appropriate cardinality. These I shall call secondary aggregates.

If from such a secondary aggregate one particle is missing, the rest will not be a possible interpretation for a RS, and the distinction of the remaining particles will be no more than momentarily in abeyance; that is, such a defective aggregate lacks verifiable properties. Likewise if there were an extra particle: if the case were possible at all, we would once again have an unstable aggregate. Though not impossible, it is unlikely that a specific number of particles should have a unique stability of this kind; it is is more plausible to assume that the distinguishing feature of a secondary aggregate should be a limiting case of the instability afflicting smaller collections. That is to say, our secondary aggregates ought to represent a threshold at which some kind of catastrophe supervenes.

What sort of catastrophes are these? We can here draw on the first and most obvious of the correlations between the cardinals of RSs and physical fact. Those of $\underline{D2}_3$ and $\underline{D2}^*_4$ are 137 and $17_{10}37$, which are near enough in the ratio of the coupling constants of the electromagnetic and gravitational fields. The former is very close to the reciprocal of the EM coupling constant α which is known with some accuracy. We are therefore strongly led to associate these two RSs with these two fields respectively. If these same two numbers are cardinals of secondary aggregates, these may represent thresholds of EM and gravitational instability respectively. And indeed it seems to be so, as has been pointed out by Noyes [4].

If $17_{10}37$ nucleons are packed within the threshold of distinction proper to such particles, which I take to be their Compton radius, we get a minimal black hole. In this case, the precise number is not very accurately known. But in the EM case, if we pack 137 electron-positron pairs inside their Compton radius, thereby removing their one ordinary distinction, their collective electrostatic potential becomes sufficient to create a further such pair thereby setting up a chain reaction which destroys the aggregate.

Pursuing the analogy, we would expect an aggregate of 10 particles associated with the strong interaction. Here we fall down, because none such is known. However, it may be that the fact that all mesons susceptible to the strong interaction are unstable gives us a getout; I shall make use of an analogous argument in another context in Chapter XII.

This is that when particles are unstable, they may function as aggregates themselves; that is, their own constitution becomes relevant. I therefore leave secondary for primary aggregates, till this matter has been explained further.

11.D. PARTONS

If we may assume that the Weak Interaction can be unified in a gauge theory with the EM field, there are four interaction fields at present known to physics, which may reflect constraints from the inchoative plane. For the gravitational and EM fields, the respective coupling constants are proportional to the reciprocals of the cardinals of the associated threshold aggregates, and may therefore directly depend on the RSs $D2_3$ and $D2_4^*$. The status of the Strong Interaction is less clear, but might yet prove to be similar. The fourth of these interactions is that described as Chromodynamic; this differs sharply from the others, in that it is developed to explain why quarks, assumed to be of a corpuscular nature, have never been observed outside of their recognised aggregates as mesons and baryons.

Now my primary aggregates of indistinguishables differ from secondary ones in just this way. Being directly composed of indistinguishables (which, being unlabellable, cannot occupy any fixed states, and so cannot become distinguished by their states, as can the immediate constituents of secondary aggregates) primary aggregates exist only with a fixed cardinality, and do not have the threshold property associated with secondary aggregates. In fact, the Hypothesis (8.1) obliges us to consider *all* observable 'elementary particles' as being aggregates of this class. The RSs available for interpretation in this way include $D2_0$, $D2_1$, $\underline{D2}_1^*$, and presumably $D3_0$ as an isomorph of $D2_1$; it may be that $\underline{D2}_2$ has also a role here, but I shall not pursue this speculation.

One is at once struck by the correspondence of the four cardinals ${}^cD2_0 = 2$, ${}^cD2_1 = 3$, ${}^cD3_0 = 3$, ${}^c\underline{D2}_1^* = 5$ with the picture presented by the parton structure ascribed to hadrons. Mesons consist of *two* quarks, hold together by exchange of gluons, of which there is one at any one moment, making *three* partons; baryons have *three* quarks, with supposedly two gluons at any one moment, making *five* partons. Making the suggested interpretation carries the implication, which may not be acceptable to all presentations of QCD, that gluons differ from quarks only in the predication for them of different colour descriptors, all the partons being essentially indistinguishable. As we shall see, the explanatory power of this interpretation is considerable, though in view of the small numbers involved its successes might be fortuitous.

Moreover, it involves a further consequence, which may be even less palatable to conventional theories. This is that leptons have the same constitution in terms of partons which is accepted for hadrons though with different flavour descriptors. It is the general opinion at present that leptons are themselves analogous to quarks, and thus radically unlike hadrons; it will be plain that, whatever the truth may be, the present theory cannot be reconciled with this opinion, for it holds that any body even indirectly observable by itself must be an aggregate of indistinguishables (though these need not necessarily be thought of as objects: they might equally be seen as descriptors). This, then, is a point of likely contention.

11.E. A RUDIMENTARY CHROMODYNAMICS

This is an appropriate place to reiterate that the present theory is not, in itself, a physical but rather an inchoative theory. It predicts certain constraints on what may and may not exist on the physical plane, and thereby may guide us (if it is accepted as true in its own field) in choosing between rival physical theories, and perhaps in constructing new ones. In regard for example to the QCD theory the above conclusions afford it a somewhat equivocal support, with certain implications which may not be generally acceptable. I have already mentioned one, the composite character ascribed to leptons. A second possible disagreement arises from the colour descriptors assigned to the several partons, which I shall now look into.

The colour descriptors used in conventional QCD are three; or rather, have three degrees of freedom, giving eight values in all. The three colours are called red, yellow, and blue; in stable aggregates, they must collectively make up 'white' understood as the sum of these three. In addition, there are three complementary colours, denoted by $\bar{r}$ (green), $\bar{y}$ (violet), and $\bar{b}$ (orange); and complementary to white is $\bar{w}$ (black). The quarks which carry these colours are supposed to be able to change them by exchanging gluons, which have other colours. Thus, a 'red' quark, by emitting a gluon bearing the $r\bar{y}$ colour description, becomes 'yellow', while a yellow one, on receiving the $r\bar{y}$ gluon, turns red. These exchanges, continually in operation, generate an attractive force between the quarks, in the same manner as the exchange of photons is used to account for the electrostatic field.

The point made by the present theory is that quarks and gluons differ from each only in their colour descriptors. From this point of view, it is inadmissible to assign 'primary' colours to quarks and mixtures like $r\bar{y}$ to gluons; instead, to preserve the symmetry required of indistinguishables (as we believe the partons to be), at least 4 different colours must be employed (to span the colour octet omitting white and black), and the sum of all those in existence in a given aggregate at a given time must equal white (or black). Schemata conforming to these two rules are shown below.

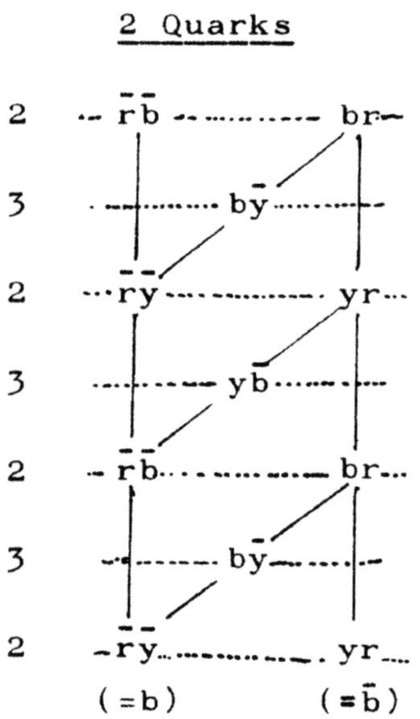

(11.3)

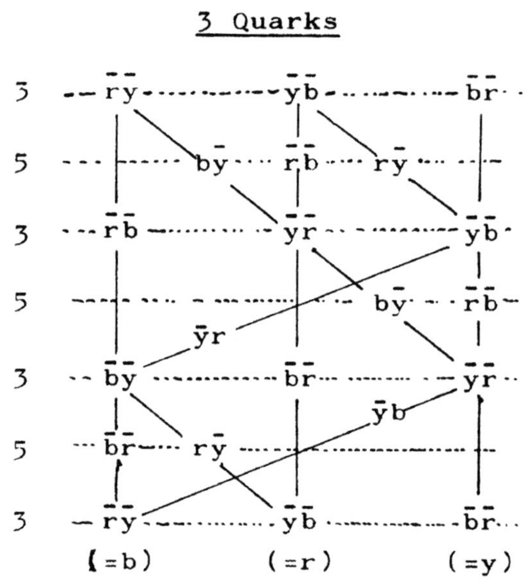

(11.4)

In (11.3), we have the case of an aggregate of two quarks of complementary colours (adding therefore to 'white'). The one in the left-hand column would conventionally be designated as 'yellow' (neither red nor blue), and that in the right-hand column 'violet' (blue + red). By exchanging a $\bar{y}b$ gluon the latter loses a b and *gains* a y, becoming (red + yellow) 'orange', while the left-hand one loses its b and gains a $\bar{y}$ thereby becoming (neither yellow nor red) 'blue'. While there are three partons in existence their colours are e.g. $\bar{r}\bar{b} + \bar{y}b + ry = w$. There are six colours altogether, in three complementary pairs. That there is no essential difference between the colour descriptions attached to quarks and gluons is shown by interchanging by b with $\bar{b}$ throughout (11.3); this makes no observable difference, since the aggregate as a whole is still always 'white'. The schema has two further twin versions, in which the colour omitted in the left-hand column, instead of being red as shown, is either blue or yellow. The whole has, apart from these variants, just two degrees of freedom: we may choose one of the quarks and we have then two choices for the first gluon it emits, and the rest follows.

In (11.4) the main differences are (a) symmetry is maintained by the gluons circulating not reversing; (b) in any given version only five colours appear; (c) with given quark descriptors there are three ways of assigning the gluon-descriptors; (d) there are eight ways of assigning quark descriptors, interchanging each or all of b, r, y with their complements; (e) the system has three degrees of freedom.

11.F. INTERPRETATION

The foregoing wholly qualitative account of 'colour' in parton systems makes use, for expository purposes, of the edges of the colour cube rather than the vertices; but since not more than six edges are used together, they can be mapped onto a set of vertices which will in all cases include the three primary colours. If each of the systems illustrated is understood as a particle, then each exhibits two alternative aspects; in one aspect, it is composed of quarks alone, and in the other of quarks and gluons. There is no other distinction involved between these two classes of partons; in particular, the degrees of freedom in each system must be exercised over both classes.

Although I have used the term 'emit' in relation to the gluons, this cannot be understood in the ordinary dynamic sense; it does no more than name the transition from the quark state to the parton state. The partons, it must be remembered, are not correlatives of indistinguishables, but indistinguishables themselves. They cannot be thought of therefore as real objects; we have no warrant, within the present theory, for supposing them anything but bearers of descriptors, which will be discussed in the next section. Nevertheless, if they are thought of as objects, it is to be expected that a coherent mathematical theory of their interactions could be constructed; for the schemata (11.3), (11.4) are closely analogous to those which portray the interaction of more substantial objects such as electrons and protons. This is the view the present theory takes of the Yang–Mills theory and later developments of QCD; it imposes an instrumentalist interpretation on QCD – but I immediately admit that my own theory is equally if not better suited to that role. In this context, at this level, reality is not a predicate to which theories can lay rival claims.

To return to our RSs: it is obvious that the dual aspect of whatever it is that the above schemata represent makes it possible to refer the two of them to four RSs. (11.3) corresponds to $D2_0$ and $D2_1$, and (11.4) to $D3_0$ and $\underline{D2}_1^*$. Note further that although the first two are both perfect Sorts, $D2_0$ is a subsort of $D2_1$, as the two quarks are of the three partons; and $\underline{D2}_1^*$ is a mixed Sort which can be expressed equally as $\cup\, D2_1\, D2_0$ and as $\cup\, D3_0\, D2_0$. The extra distinction in the latter case between the Subsorts is perhaps reflected in the fact that in (11.3), but *not* in (11.4), the colours of the gluons sum to white like those of the quarks; but this is a collective property, and does not serve to identify any one gluon of (11.4) as such.

All this amounts a quite detailed correspondence between the four RSs listed above, and the currently assumed constitution of mesons and baryons respectively. The quantitative theory is inevitably missing, however, so that the weight the correspondence contributes to the assessment of the total theory is no more than the numerological agreement with the figures 2:3:3:5.

11.G. PARTON DESCRIPTORS

This then completes the search for correlatives of the RSs of the family D2 in term of aggregates. I now turn to their alternative role as components of descriptors. It is not to be expected that anything will have suggested itself as requiring 137 descriptor-components; such objects are of course a commonplace in biology, but are not expected in physics. I offer no conjectures as to whether such things exist or not (still less on the reality of the next class); if they do, it is small wonder they have been overlooked. But we ought to be able to find things requiring the more modest numbers of 2, 3, and 10 components to complete their description.

$D2_1$ with its three members can obviously be correlated with the three primary colours of the colour octet assigned as descriptors to the partons of the previous section. This would be a baseless conjecture, if they were not associated with a further 7 components, which are (at present) collectively required to specify the 'flavour' of a quarks. This is the cardinal of $D2_2$, not itself an RS but a perfect Subsort of $\underline{D2}_2$: it is so far unique as the only perfect but non-rational Sort having a plausible correlative already named. 'Colour' and 'flavour' together make up the 10 components of a possibly complete quark description; according to my interpretation, such a description applies to any parton, since there is no necessary distinction between quarks and gluons.

This conclusion is however a very brittle one. It implies that there are no more than 7 quantum numbers (or rather, 7 degrees of freedom in assigning quantum numbers), which can still be believed; but in view of the accelerating rate at which they have been being discovered the number 7 is facing imminent supersession.

There is no corresponding threat in the steady increase in the number of quark species claimed; our 7 descriptor-components set no near limit on this number. On the other hand, we are committed to accounting not only for mesons and baryons, but for leptons also; and for these some kind of quark-analogues must be postulated to exist. The table (11.6) below gives the descriptors (quantum numbers) conventionally ascribed to the four best-known quarks, alongside those which I propose for four corresponding 'larks'; from these, using the admittedly *ad hoc* selection rule, that $-1 < (M + C) < 2$, we can construct all the well-established leptons (some appearing as their antiparticles) and one which is not at present known. There is a considerable degree of symmetry between the half-tables for quarks and for larks, as one might expect; but the latter remain merely speculative at present.

Quantum Numbers		Quarks				Larks			
		u	d	s	c	u^*	d^*	s^*	c^*
Q	Charge	$+\frac{2}{3}$	$-\frac{1}{3}$	$-\frac{1}{3}$	$+\frac{2}{3}$	$+\frac{2}{3}$	$-\frac{1}{3}$	$-\frac{1}{3}$	$+\frac{2}{3}$
J	Spin	$\frac{1}{2}$	$\frac{1}{2}$	$\frac{1}{2}$	$\frac{1}{2}$	$\frac{1}{2}$	$\frac{1}{2}$	$\frac{1}{2}$	$\frac{1}{2}$
I	Isospin	$\frac{1}{2}$	$\frac{1}{2}$	0	0	0	0	0	0
Y	Hypercharge	$+\frac{1}{3}$	$+\frac{1}{3}$	$-\frac{2}{3}$	$+\frac{1}{3}$	0	0	0	0
M	Muon Number	0	0	0	0	$-\frac{1}{3}$	$-\frac{1}{3}$	$+\frac{2}{3}$	$-\frac{1}{3}$
B	Baryon Number	$+\frac{1}{3}$	$+\frac{1}{3}$	$+\frac{1}{3}$	$+\frac{1}{3}$	0	0	0	0
L	Lepton Number	0	0	0	0	$+\frac{1}{3}$	$+\frac{1}{3}$	$+\frac{1}{3}$	$+\frac{1}{3}$
C	Charm	0	0	0	$+1$	0	0	0	$+2$

(11.6)

This table can be regarded as an 8×8 matrix. As such, it is singular, since $2J - 3B - 3L = 0$; the 'rank' of the matrix is in fact not 8 but 7, indicating that it contains seven degrees of freedom. It is these, together with the three primary colours of the chromodynamic system, that make up the 10 components of complete quark (and particle) descriptors.

As for the leptons, there are eight combinations of the four 'larks' for which M + C is either 0 or 1 (and a further eight with reversed signs), which give the following equivalences:

$$\begin{array}{lll} d^*d^*s^* = \varepsilon & d^*s^*s^* = \mu & \\ u^*d^*s^* = \nu_\varepsilon & u^*s^*s^* = \nu_\mu & \\ u^*u^*s^* = \bar{\varepsilon} & & u^*d^*c^* = \bar{\tau} \\ & & d^*d^*c^* = \bar{\nu}_\tau \end{array} \tag{11.7}$$

The eighth combination is $u^*u^*c^*$ which ought to be an anti-tauon of charge + 2. Analogously with the mesons, $u^*\bar{u}^*$&c would give photons. But, as I have said, this is speculation; nothing decisively confirms the correctness or otherwise of the right half of (11.6). The calculation in Section 12.C requires only that leptons should have five partons, but says nothing about their individual descriptors.

11.H. RECONSTRUCTION OF PARTICLES

One of the results of my identifying partons as the proper 'physical' entities to correlate with some of our smaller RSs has been to diminish the importance attached to the distinction made in QCD between quarks and gluons; I take these names to refer to different values, taken from one class, and predicated of indistinguishables. Thus whereas the prevailing theory has it that when we bring three quarks together, having different 'colours', they will interact by exchanging gluons, I put the same thing differently, saying that any three partons which between them have descriptors uniquely specifying what sort of fermion they belong to will acquire two companions to make up the required fivesome. Likewise for bosons, whose three partons can be sufficiently specified by two.

This does not amount, on either view, to a regular physical process; as regards stable particles, it is a mere theoretical position. But the situation where a three and a five, or a two and a three, are equivalent may be important for particles which are unstable, at least at the moment of their decay. It might show up statistically in such conditions as those prevailing at the birth of the universe (what has been referred to by some as the 'quark soup' phase), or in aggregates of unstable particles. It is to the determination of the size of the threshold aggregate characterizing the strong interaction that I shall apply these ideas first; the cosmogonic relevance of the process will be considered in Section 12.G below.

The problem then is to assemble 10 indistinguishable pions, given that they come in two charge states (neutral pions are very short-lived and I shall ignore them); if the charges are equally balanced, we shall need 20 pions to get 10 indistinguishables, or if they are randomly assigned about 18 would give the required expectation. But if the aggregate is in the condition requisite for 'reconstruction' to occur, these numbers could be attained in the presence of only 40 (or 36) partons, which could be provided by no more than 14 (or 12) input pions. Taking account of the fact that the distribution of parton descriptors is not likely to be optimal, these numbers should be slightly raised – 14 is still enough, but the lower estimate might be nearer 13 than 12.

All this is disappointingly vague; it is a statistical estimate based on a rather crude picture of what might be going on. But since the result is within range of the accepted figure, and since moreover the same argument applied in another context in Section 12.G also makes some kind of sense, it is perhaps worth including among our conclusions.

11.J. SUMMARY OF KINDS OF PARTICLES

The conclusions which we have arrived at in this chapter can be briefly summarized as follows:

> All 'particles' which are individually observable are secondary indistinguishables, that it, aggregates of primary ones (11.8.1) C.
>
> They are of two classes: one composed of three partons with two degrees of freedom; and one of five partons, with three degrees of freedom (11.8.2) C.
>
> They interact with three different external fields, characterized by different stability-thresholds, namely
>
> for gravitation, a threshold aggregate of $17_{10}37$ nucleons (11.8.3) C.
>
> for electromagnetism, a threshold aggregate of 137 electron-positron pairs (11.8.4) C.
>
> for the strong* interaction, a threshold aggregate of perhaps 13 or 14 pions (11.8.5) C.

* This assumes that the RS $\underline{D2}_2$ is correctly correlated with the strong interaction; but $\underline{D2}^*_1$ is also a candidate for an even stronger interaction (coupling constant 0.2 rather than 0.1) which may or may not prove to be involved.

CHAPTER XII

STATES OF PARTICLES

We are evidently far from having accounted for all the kinds of particles at present known by the physical correlates proposed in the last chapter for the rational Sorts in the family D2. To explain the masses, modes of decay, lifetimes, and other properties of all the two hundred or so species which have been discovered is manifestly beyond the competence of my theory; at best it may make some of the assumptions incorporated in Quantum Chromodynamics seem a little less arbitrary than they now do. If so, it will have added to the already formidable explanatory power of that theory. But we have now no further rational Sorts to examine, from which additions to this repertory might emerge, and our journey is nearing its end.

We have, however, still to look into the family D3, and we shall find here Sorts which seem to be reflected in a system of particle-states, which will be of some interest. There are one or two other points to clear up under this same heading. The remarks made in Chapter X on the 'states' (or rather, the lack of any definable states) available to entities when separated by very small distances, requiring the use of 'disordinate statistics', can now be used to calculate quantitative results which are well confirmed. I shall conclude this chapter with an attempt to enumerate the totality of states which can be used to distinguish members of one class of secondary indistinguishables, the nucleons; as we shall see, we don't win every time.

12.A. THE FAMILY D3

This family has only two rational Sorts, $D3_0$ and $\underline{D3}^*_1$. The former has already received two interpretations, perhaps three. As one of the initial Sorts, it has been correlated with the dimensionality of space; and again as the Boolean lattice B^3 realized in the colour octet of the Yang–Mills theory of quarks – which may perhaps be regarded as a special case of the same, though there is no sense in thinking of the colour octet as occupying real space. In its alternative capacity as one of the family D2, $D3_0$ has also been correlated with the three quarks constituting a baryon (but I say 'fermion').

The three spatial coordinates are themselves involved in describing the state of a particle, for where it is, relative to others, is certainly part of what we mean by its state, however inconstant a part. But it is $\underline{D3}_1^*$, of which of course $D3_0$ is a Subsort, which we are particularly concerned with in this connection. This Sort has 10 members, and we must look therefore for some set of ten descriptors of which might help to locate a particle in a 10-dimensional 'space'.

To fully describe an event, as part of a 'causal' nexus in space-time, we need to specify the trajectories of all its incident and emergent particles. The theory requires that field strengths are fully determined *without* reference to intermediate points of W by those at all the events involved. The data needed for any one particle are therefore: the event location (4 components), the momentum or velocity of the particle (3 components), and the force or acceleration acting on it (3 components). These components being in general mutually independent, we have 10 degrees of freedom. The information can be classified in other ways, but always with the same liberty. This corresponds with that of the ten potentially independent components of the energy-momentum tensor, occurring in a corresponding role in relativistic quantum-mechanics. The argument is clear enough to justify the statement that:

> We correlate $\underline{D3}_1^*$ with the degrees of freedom of the energy-momentum tensor specifying the state of a particle; and we predicate of them the values of its ten components applied to a given particle (12.1) C.

At the present point in the argument we have enough of the picture filled in to be able to say with some confidence that the theory *predicts* that the state of a particle (in the above sense) must have 10 components. This prediction is fulfilled. We cannot predict the tensor itself as a means of presenting these components, because that belongs to physical theory. But there is here an element of wonder not previously encountered, since our '10' is derived as $(2^2 - 1) + (2^{2^2 - 1} - 1)$, whereas the physicists '10' is $\Delta 4 = 4.5/2$, the degrees of freedom of a symmetrical tensor in 4-space! This is, to say the least, an odd coincidence; on the nature of which the reader may like to ponder.

There is another point here. There being no rational Sorts between $D3_0$ and $\underline{D3}_1^*$, we do not expect any physical reality to attach to the subspaces of 4 to 9 dimensions, analogous to those which we do have physical correlates of in the case of the 3-space attached to $D3_0$. Their absence accords well with the fact that no meaning can be given to an *incomplete* tensor. Also, this is the best case in which we can interpret the (non-rational) perfect Subsorts of $\underline{D3}_1^*$. Three concern location, and seven concern motion, and the nature of the problem does assign distinctly different roles to these two classes of descriptor. Altogether this result strikes one as more evidential than the smallness of the number 10 would by itself allow.

12.B. STATES IN A DISORDINATE SPACE

I now revert to a consideration of what happens in those regions of space where there are less than two events, so that the disordinacy ascribed to the space V becomes relevant. In a wider environment, particles are in states understood to be defined by space-time vectors, and their behaviour is governed by the dynamics associated with such states. In view of (10.2.3) we clearly lose touch with the familiar space-time geometry; but how? Conventional treatments assume that intervals retain the distinction between space-like and time-like proper to more familiar environments, which is evidently not so. But that measures of 'distance' established in the presence of such distinctions can be, at least up to a point, extrapolated into such small regions, is presupposed by my contention that a meaning can be attached to thresholds of indistinguishability, which are measures of distance even if limiting ones. Such distances are in practice defined by the assumption that dynamical principles, and in particular the inverse-square laws belonging to the extended fields of electromagnetism and gravitation, continue to hold down to the threshold of distinction for the present theory, but not down to zero separation as they do 'classically'.

Under these circumstances, we have to apply the disordinate statistics for a 3-space, as defined in (10.B). To illustrate what happens, I shall now apply these statistics to evaluating the self-potential of a particle subject an inverse-square field, which involves a region so small that no exact locations exist. As we have already seen in Section 11.C any particle, to be 'observable' at all, must represent a rational Sort, that is, an aggregate of indistinguishable entities, among which its 'weight' (charge or mass) must be apportioned, each being at one point of V, but referred (without specification of coordinates) to the intuitive space W for dynamical purposes. These fractional 'weights', being associated with unobservables, cannot be individually quantified, but may take any values provided they sum to the value proper to the particle in question, which of course *is* observable

For any dipole, or pair of fractional 'weights', the potential is given by

$$u = w/v = x(s - x)/v \tag{12.2}$$

where s is the total weight of the dipole, x the fraction associated with one pole, and v is the distance between the poles. Since it is understood that this is a spatial and not a temporal measure, the statistics to be applied are those for a 3-space, notwithstanding that we can no longer ask how this is related 'geometrically' to the total 4-space of which it is a subspace.

From (12.2) we see that $u = wv^{-1}$, and since w and v^{-1} are independent variates we can write the expectation of u, $\langle u \rangle$, as $\langle u \rangle = \langle w \rangle \langle v^{-1} \rangle$, allowing us to consider separately the probability distributions for w and v^{-1}. I shall look first at w.

The mean value of variate w over a range $(0, a)$ must be proportional to a by the principle of dimensions, which are the same for a and w. But the mean $\langle w \rangle = \int_0^a w\mathrm{P}(w)dw / \int_0^a \mathrm{P}(w)dw$, and if we write this as $a\mathrm{X}$, $\langle w \rangle \propto a$ only if $d\mathrm{X}/da = 0$. If $\mathrm{P}(w) \propto w^p + w^{p+q}$, this condition entails that $q = 0$: from this beginning we can generalize over all real functions capable of representing $\mathrm{P}(w)$, and conclude that we must have $\mathrm{P}(w) = hw^p$, where p is a real number other than -1 or -2. Now in our case $w = x(s - x)$; s and x are independently variable, so that $\mathrm{P}(w) = \mathrm{P}(x)\mathrm{P}(s - x)$ with the second x restricted to be equal to the first. Now since x by itself can in principle take both positive and negative values, whereas $\mathrm{P}(w)$ must be positive real, we must write $\mathrm{P}(w) = hw^{2n}$ where $n \neq -1$. I shall appeal to the principle of ignorance (8.4) to choose the value $n = 1$, and conclude that

$$\mathrm{P}(w) \propto w^2 \tag{12.3.1}$$

As for v^{-1}, we note that v is a distance measure, and consequently governed by the rule (10.2). Since we have here to consider W as a 3-space, and q in (10.16) $\equiv -1$, we find that $p = -5$, so that

$$\mathrm{P}(v^{-1}) \propto v^{-5} \tag{12.3.2}$$

We have now to find a formula for the expectation $\langle u \rangle$ taken over both variates w, v^{-1} within the appropriate limits. For w, since $x \leqq s$, we may take $0 < w < s$; and for v the limits must be $d < v < \infty$, where d is the appropriate threshold of distinction. We have therefore, for given s, that w is a function only of x, and

$$\langle u \rangle = \int_0^s w \int_d^\infty v^{-1}\,\mathrm{P}(v^{-1}).\,dv.\,\mathrm{P}(w)dx \bigg/ \int_0^s \int_d^\infty \mathrm{P}(v^{-1})\mathrm{P}(w).\,dx.\,dv \tag{12.4}$$

On integrating first by dv, after setting $P(v^{-1}) = v^{-5}$, we get $\int_d^\infty v^{-6}dv / \int_d^\infty v^{-5}dv = 4/5d$. Then, setting $w = x(s-x)$, $P(w) = w^2$

$$\langle u \rangle = \frac{4}{5d} \cdot \int_0^s x(s-x)x^2(s-x)^2\,dx \Big/ \int_0^s x^2(s-x)^2\,dx$$

$$= \frac{4}{5d} \cdot K_1 \,.\, s^2 = \bar{u}_1 \quad \text{where} \quad K_1 = \frac{3}{14} \tag{12.5}$$

In general, of course, there will be more than just two fractional 'weights' to consider. Suppose there are n degrees of freedom: we can then consider the effect of one fraction against the rest. The total, ignoring the common multiplier $4/5d$, is from (12.5) $K_n s^2$, the fraction of weight y contributes $y(s-y)$, the rest $K_{n-1}(s-y)^2$ and so

$$K_n s^2 = \int_0^s [y^3(s-y)^3 + K_{n-1}y^2(s-y)^4]\,dy \Big/ \int_0^s y^2(s-y)^2\,dy \tag{12.6}$$

which evaluates to

$$K_n = \frac{3}{14} + \frac{2}{7}K_{n-1} = \frac{3}{14}\sum_{i=0}^{n-1}(2/7)^i \tag{12.7}$$

giving for the first three members of the series

$$K_1 = 0.214245715; \quad K_2 = 0.275510204; \quad K_3 = 0.293002915$$

and

$$\bar{u} = \frac{4}{5d} \cdot K_n s^2 \tag{12.7a}$$

The values to be substituted for d, n, and s depend of course on what field and what kind of particles we have to deal with.

12.C. THE PROTON-ELECTRON MASS RATIO

The above result can be tested by applying it to the electron, on the assumption that its mass, as has often been suggested, is wholly due to the intrinsic electrostatic potential of its charge. The arguments given in Chapter XI require us to attribute to the electron, as an 'observable' particle, a constitution involving five 'partons', analogous to that of hadrons with three quarks and two gluons; this is admittedly an unorthodox opinion, and is in need of support, which is partly afforded by the present calculation. Among these partons there are two perfect Subsorts, and therefore *two* constraints on their fractional charges (one of which is that all together sum to e) leaving us with three degrees of freedom, so we take $n = 3$. Obviously, we have also to set $s = e$.

As for the threshold of distinction d, we can have no direct evidence of its magnitude, the partons being unobservable of themselves. Presumably, again following Chapter XI, the value required is the Compton radius of some particle. Obviously not that of the electron itself, within which many electrons can be assembled, but something much smaller. I propose to take for d the CR of the proton (the most massive stable particle, and hence the smallest CR available), that is to say, $d = h/2m_p c$ where m_p is the proton mass. We have therefore, substituting into (12.7a),

$$m_e c^2 = \bar{u} = \frac{4}{5d} \cdot \mathrm{K}_3 \cdot s^2$$

$$= \frac{4}{5} 0.293002915 \; e^2 \, 2m_p c/h$$

and setting $e^2 = \alpha hc/2\pi$ this reduces to

$$m_e c^2 = 0.234402332 \, \alpha m_p c^2/\pi \tag{12.8}$$

Now the value empirically established for the constant α is $1/137.036$. This figure, however, goes with a different theory from ours, and for consistency we ought to substitute the value $1/137$ being that given by the association of the rational Sort $\underline{\mathrm{D2}_3^*}$ with the electromagnetic field. This leads to

$$m_p/m_e = 137\pi/0.234402332 = 1836.1515 \qquad \text{(12.9) C.}$$

which agrees with the empirical value to well within the experimental error estimated as 0.00070. This gives some support to the admittedly unusual model of the electron assumed above.

12.Ca. THE FINE STRUCTURE CONSTANT

The electron-proton mass ratio is derived above by a calculation taking literally the 'disordinacy' attributed by my theory to the space V. The properties of empirical space do not enter this calculation, because what we have in mind is a ratio of *masses* – which we may presume to be independent of spatial measurements. In this calculation I used the theoretical value of $1/137$ for the fine structure constant. As is well known, the empirical value is slightly less: I propose that this arises because the constant is empirically determined as a ratio of quantities involving space irreducibly, and the properties of space underlying the required measurements are *not* those of V, which has no *measurable* extent at all.

As argued at length in Chapter X, measurements of length presuppose the space there called W, which is notionally of a piece with the 'space' of man-scale observations. Its reality at very small scales, such as we are here concerned with, is highly dubious; it is extended into such applications by a simple-minded extrapolation of the laws of dynamics from contexts in which these are well-established into the unknown. I propose to assume however that any deficiency in this procedure can be corrected by going over from W to V whenever we encounter situations involving less than two events. The uncertainty principle requires that we observe at least two events involving one electron if we are to determine *accurately* the ratio we define as the fine-structure constant. It should therefore be legitimate to carry out the calculations required entirely within W, provided we take care that the relevant comparisons with the situation in V are well-founded.

The origin of the correction to be applied to the theoretical value 1/137 is evidently that my assumption that the electron is of infinite extent (as it must be in V, where infinity is only next-door, so to say) is not correct in W, where two events are always separated by a finite interval. We should, therefore, replace the upper limit of integration with respect to v in (12.4) by a finite distance which we expect to emerge from an analysis of the electron self-potential in W.

We begin then by considering the geometry of the electron in W. The naive picture of it as a spherical hypercylinder extended in time is wrong on two counts: there are no grounds for believing it to be continuous in time; and its spin symmetry requires that its 3-dimensional connectivity is toroidal. I therefore assume that it can be represented as the hypersurface of a hypertorus; we have already encountered this figure as a possible one for the cosmos as a whole, and there is no other form in which the torus can be given a finite extension in a fourth dimension, so as to be compatible with the symmetry required in the space V. The discrete partitioning of the charge among five partons of course does not hold in W, and we have to consider what to replace it with.

One realization of the hypersurface is as an indefinitely repeated cubic lattice, so we might consider point charges in such an array; each charge will be surrounded by a box of 26 nearest-neighbours, and the potential can be analysed into dipoles of which 6 will have minimal lengths, 12 will be $\sqrt{2}$ times as long, as 8 will be $\sqrt{3}$ times as long. All our calculations assume however, for the sake of isotropy, that we take our measurements in an equiaxial hypertorus, and one might suppose that these ratios would no longer hold in that model. That they do can be illustrated by the following diagram ,which shows, using 2 dimensions instead of 3, an equiaxial torus on the left (which can be seen as a double-skinned sphere) and the corresponding square lattice on the right:

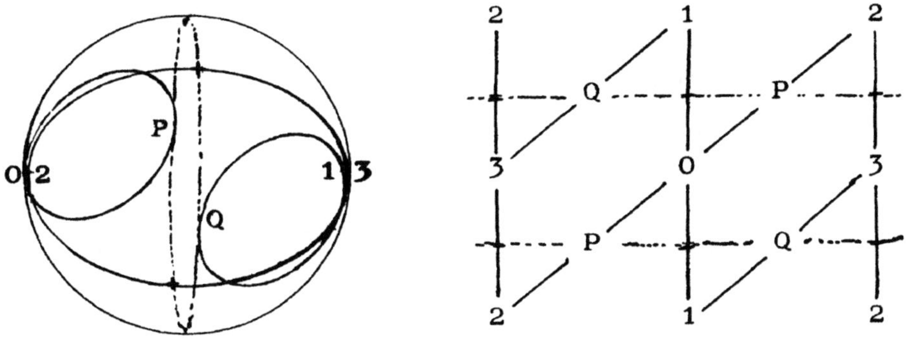

The two poles of the torus are P, Q, and 0, 1, 2, 3 are a set of four mutually antipodal points on the equator. The shortest route from 0 to 2 (which coincide in the diagram) is the smallest circle through a pole, of which that through P is indicated, and also that from 1 via Q to 3. In the corresponding projections onto the square lattice it is clear that the distance 0, 2 is $\sqrt{2}$ times 0, 1; the former corresponds to the circumference of the circle 0, P, 2, which is easily seen to be 1/2 the equatorial semicircumference 0, 1. The 3-dimensional case cannot be illustrated, but provides similarly for the $\sqrt{3}$ ratio.

We therefore visualize the potential as the sum of that from 26 dipoles in such an array, each contributing $(e/2)^2/s$ where s is the length in appropriate units. This length is that of the semicircumference from 0 to 1 in the diagram, namely πr if r is the radius of the hypertorus. However, we intend eventually to refer the result back to a hyperspherical model, and whereas the hypersurface of a hypersphere is $\pi^2 r^3/2$, that of a hypertorus is $4\pi^2 r^3$, so that the radius of the hypertorus must be taken as $r/2$ in terms of the hyperspherical radius. We accordingly take the dipole lengths to be $s_1 = \pi r/2$, $s_2 = \pi r/\sqrt{2}$, $s_3 = \pi r\sqrt{3}/2$ where r is the radius of the hypersphere. The total potential is therefore the sum of 6, 12, and 8 of the respective values of $(e/2)^2 s$, divided further by 3 to allow for the three identical and therefore *non*-additive identifications of the hypertoroidal axes with arbitrary hyperspherical ones. Thus

$$p = (e/2)^2 . \left(6 + \frac{12}{\sqrt{2}} + \frac{3}{\sqrt{3}}\right)\frac{2}{\pi}\bigg/ 3r \qquad (12.10)$$

$$= 1.013563108\, e^2/r$$

which we must identify, by finding the right value of r, with the known electron mass $m_e c^2$. We use the equation $e^2 = \alpha hc$, and setting $\alpha = 1/137$ (the correction being calculated can be shown *a posteriori* to be negligible in this context), so that

$$r = 1.013563108 . h/137 m_e c = 4.2848137_{10} - 13 \text{ cm.} \qquad (12.11)$$

$$= 2.16182\, d$$

where d is as before the Compton radius of the proton.

Up to this point I have assumed a wholly static configuration. The internal momenta among the half-charges have been assumed precisely zero, which entails that the linear dimensions of the dipoles are wholly unknown, by the uncertainty principle. Suppose then that one end of a pair of dipoles is displaced, so that one path is increased from s to $s + x$ while the other is diminished to $s - x$. The resulting potential will be not $2e^2/4s$ but

$$\bar{p} = \frac{e^2}{4}\left(\frac{1}{s + x} + \frac{1}{s - x}\right) = 2e^2 s/4(s^2 - x^2),$$

and its corresponding probability is proportional to the time the system spends in this state; this, having the dimensions of action, is quantized and so in our context constant, so the required weighting function is $(s^2 - x^2)/s$, and the mean potential is

$$\bar{p} = \frac{e^2}{2s} \cdot t(s)$$

$$= \frac{e^2}{2s} \cdot \int_0^{s-d} dx \bigg/ \int_0^{s-d} (s^2 - x^2) dx \qquad (12.12)$$

assuming that, as before, the charges approach no nearer than d. On evaluating the integral, we find that

$$t(s) = 3s^2/(2s^2 + 2sd - d^2) \qquad (12.12a)$$

This is the ratio by which the potential of a pair of dipoles of mean length s is enhanced by the uncertainty of position; the effect considered is the tangential one, and we have yet to take account of the radial component of the uncertainty. Each corresponding length must be increased in the same proportion to preserve the value of $\bar{p}$; thus, when $s = \pi r$ it will become $\pi r \cdot t(\pi r)$, while $s = \pi r\sqrt{2}$ becomes $\pi r\sqrt{2} \cdot t(\pi r\sqrt{2})$, etc. The effect on r is therefore

$$\bar{r} = r(3t(\pi r) + 6t(\pi r\sqrt{2})\sqrt{2} + 4t(\pi r\sqrt{3})\sqrt{3})/13 \qquad (12.13)$$

which can be calculated from the formula for $t(s)$, in units of d, as

$$\bar{r} = 1.99188r = 4.306085\,d \qquad (12.13a)$$

Turning now to the radial component of the uncertainty, we note that for any given value assigned to r, the potential over the whole hyperspherical surface (which is the model of the electron appropriate to the space W without consideration of V) is proportional to k/r; as before, the probability of any such value is inversely proportional thereto, so that

$$\bar{\bar{p}} = \int_1^{\mathrm{D}} \left(\frac{k}{r}\right) r\,dr \Big/ \int_1^{\mathrm{D}} r\,dr = 2k/(\mathrm{D}+1) \qquad (12.14)$$

If, as before, we let the upper limit D tend to ∞, we get zero potential, so we have a finite value for D; the lower limit 1 stands for our former d, since this is now the unit of measurement for r. We have $\bar{\bar{p}} = k/\bar{r}$, so that $\bar{r} = (\mathrm{D}+1)/2$ and from (12.13a)

$$\mathrm{D} = 2\bar{r} - 1 = 7.61217\,d \qquad (12.14a)$$

On substituting this upper limit of integration into (12.4) in place of ∞, we find on inspection that if the former value for the m_p/m_e is to remain unchanged, we must multiply α by the ratio $\mathrm{R} = (\mathrm{D}^5 - \mathrm{D})/(\mathrm{D}^5 - 1)$. From (12.14a), we find that

$$\frac{\mathrm{D}^5 - \mathrm{D}}{\mathrm{D}^5 - 1} = 1/1.0002591 \qquad (12.15)$$

whence, multiplying by 1/137, we find that

$$\alpha = 1/137.0355 \qquad (12.16)$$

which departs by about 5 st. dev. from the empirical value of 1/137.03604–0.00011. A possible source of error, which would be of the right sign, might be the assumption of point charges (which in reality as here conceived are undefinable) as against some kind of disordinate dispersion.

12.D. STATES DISTINGUISHING PARTICLES

We have seen, in Section 1.B, that all elementary particles of a given kind are 'secondary indistinguishables', in that, whenever we observe two of them rather than one, we may infer that they are in different states. If, therefore, we could enumerate the states which one kind of particle could occupy, we could conclude that no more than that number of particles of the given kind could ever be observed 'simultaneously'. Of course, since simultaneity is indefinable over large regions of space (and since spatial separation is obviously one of the classes of states which would be involved), this would not be a feasible experiment. But it could still be an indirect inference from possible observations.

There seems to be only one kind of particle for which the enumeration of states could be carried out and the result at least in principle checked by observation: that is, nucleons. Since these particles account for almost all the particulate mass in the universe, an estimate of their total number could be verified by determining the total mass of the universe; and this in turn might be inferred from observations of the mean density, and estimates of its volume. Nucleons comprise protons and bound neutrons; these are essentially alternate states of one kind of particle, neutrons being without charge and 0.138% heavier than protons which have a positive charge. It makes sense to say of one nucleon that it is sometimes being a proton and sometimes a neutron, from which it follows that these states are not of the kind that could distinguish or help to enumerate the particles.

To apply this method, we have first to list all the classes of states by which two nucleons might be distinguished, and to enumerate the states in each class. The product of all these contributions gives then the total number of potentially distinguishable states in which a nucleon may be. The theory tells us that we shall never be able to observe as distinct *more* than this many nucleons. It may be that we shall observe fewer, but if so we shall want to know why. That is, information is required to account for any of the states being unoccupied. Even if we do observe the full number of particles allowed by the tally of their states, the number actually existing may still be smaller: for we cannot exclude the possibility that, because simultaneity cannot be defined over any considerable part of space, one particle may be observed in more than one state. But if that happens (as it will) there should still be enough data to enable the result to be checked by observation – if the appropriate observations can be carried out successfully.

The classes of states involved in this procedure can be listed readily enough from the interpretations already proposed for the various rational Sorts. Spatial location is one of the more obvious such states; it is associated with the Sort $D3_0$. $D2_0$ gives us the class of orientations, of which nucleons may participate in only two, defined by their opposite spins. $D1_0$ gives, if anything, only time, which evidently cannot be used for distinguishing particles. Each of the four force-fields is a potential candidate, since one way of observing particles and at least locally counting them is by means of their mutual perturbations; these we associate with $\underline{D2}_1$, $\underline{D2}_2$, $\underline{D2}_3$, and $\underline{D2}^*_4$. $\underline{D3}^*_1$ looks as if it could be used, insofar as it represents the components of energy and momentum, but this information is already subsumed in that given by gravitation. None of the other rational Sorts appear able to define the kind of states we are looking for.

12.E. LOCATION AND ORIENTATION STATES

It may appear fatuous to ask how many states are distinguished by spatial location; if the answer is not infinite, it might be such that to fill all these 'states' would be to pack all space tight with nucleons. There is, however, a class of subspaces of the intuitive space W within which spatial measurements are precluded: these are the hypercones generated by a wavefront of light incident on or emanating from a given observation-event; they are called 'light-cones'. Direct spatial measurements cannot be carried out between points within one light-cone (it is irrelevant that they can often be indirectly inferred from other considerations). Thus, all points in one light-cone are 'indistinguishable' in respect of location; these include the points where the light-cone is intersected by the world-lines of nucleons, points which the rest of our classes of states will discriminate and enumerate. The number we need to bring in at this point is the number of distinct light-cones.

It is easy to see that they are many; but even if they are finite, it is not easy to see what makes them distinct. I shall assume that the required number is equal to our largest finite cardinal, $17_{10}37$. If this leads to an even vaguely plausible result, it must be correct, for there is no other available number anywhere near it.

As for the state class depending on orientation, as I have already said nucleons are capable of only two orientations, defined by their spin. Our enumeration of states therefore goes from $17_{10}37$ to $34_{10}37$ to take account of this class. This pair of states is manifested, for example, in the magnetic moment, which may be of either sign.

12.F. INTERACTION-FIELD STATES

In seeking to evaluate the contribution to the distinction of nucleons made by the gravitational field, we pursue the same tactic as with spatial location; that is, we define a class of states within which this particular factor is eliminated, and then attach a cardinal to the class of such classes. Now a set of objects which experience no acceleration with respect to each other are said to belong to one inertial frame. On the universal scale, mutual accelerations may be caused by three factors, namely gravitational attraction, electromagnetic forces, and the universal expansion. The last is irrelevant, being from our point of view a geometrical rather than a dynamic effect; the first two can affect different particles differently, and so contribute to their distinction.

We have seen cause in (11.10.1) to associate the rational Sort $\underline{D2}_4^*$ with the gravitational field, and there is therefore good reason to postulate that the contribution of this field should enable us to distinguish $17_{10}37$ subclasses of inertial frames, each of which, on further discrimination by electromagnetic effects, should yield 137 frames, this being the cardinal of the corresponding rational Sort $\underline{D2}_3$. At this point, however, we have to consider that not all nucleons interact with this field. Almost all neutrons are incorporated in helium nuclei, which constitute perhaps 27% by weight of the matter of the universe; of this quantity, some will have been reprocessed at least once. At a rough guess, I shall therefore assume that one-eighth of all nucleons fail to respond to the EM field, thereby reducing the relevant factor from 137 to $\left(\frac{7}{8}137 + \frac{1}{8}\right) = 120$. Unlike the conjecture made in Section 12.5 on the number of distinct light-cones, which is either correct or hopelessly wrong, this guess is certainly wrong, but not far wrong.

(Some may object that, since we are calculating an integer, however large, any fraction such as 1/8 used in this way should have a denominator which divides (in this case) $137 - 1 = 136$. It is, however, sufficient that it should divide $34_{10}37$, even if the point is held to be valid. $17_{10}37$ is an abbreviation of 170141183460469231731687303715884105864.)

As for the strong interaction, this might be held to be irrelevant on the grounds that almost all nucleons are not affected by it. But this was not always so; there must have been a time when this interaction was universal (see further Chapter XIII). It should therefore be included, presumably by multiplying our total by $10 = {}^{c}\underline{D2}_2$.

We arrive then at $34_{10}37 \times 17_{10}37 \times 120 \times 10 = 6.936_{10}79$, before considering the fourth interaction-field of chromodynamics.

12.G. THE CHROMODYNAMIC CONTRIBUTION

We have seen, in Chapter XI, that there is a case for associating the rational Sort $\underline{D2}_1^*$ with the so-called 'chromodynamic' field. The corresponding cardinal is 5, and we accordingly consider every fermion to be an aggregate of 5 indistinguishable 'partons'. There is therefore, in one sense, a total of $3.468_{10}80$ partons. But $\underline{D2}_1^*$ like all the others in the D2 family is a mixed Sort, and can be written as $\cup D2_0 D2_1$ indicating a distinction, possible after one step of predication, into 3 indistinguishable 'quarks' and 2 other indistinguishables, presumptively 'gluons'.

However, once the quark descriptors are determined, those pertaining to the accompanying gluons are fixed. In this sense, every triplet of quarks generates its own gluons. This need not be a mere figure of speech (at least, no more so than anything else around here); for we have no grounds in our theory, nor precedent from analogical cases, for postulating conservation of partons or of gluons (though the accepted theory does seem to imply a conservation of quark numbers). It is therefore possible, from a given total of n partons, to build up in principle $\frac{5}{3}n$ baryons.

The objection, that such a reconstruction could not be an empirically attestable fact, carries little weight. For all along we have been dealing not with feasible experiments but with distinctions possible in principle. No one will ever actually count up all the nucleons in the universe. It is a calculation, though hopefully one capable of confirmation by indirect observation. There is therefore no reason to reject this final factor of 5/3 on this account.

If it is accepted, the number of nucleons we end up with is $\frac{5}{3}6.936_{10}79 = 1.156_{10}80$. If we assign to each a mass of 1.008 daltons, each of $1.673_{10} - 24$ gm, the total mass of the universe which is in particulate form is found to be $19.34_{10}55$ gm.

12.H. THE MEAN DENSITY OF THE UNIVERSE

The only chance of empirically testing this figure is to calculate the mean density of the universe; not that this is known with any accuracy, but there is at least some evidence on the question, and more may come to hand. We assume, as I have in general done, that the connectivity of space is that of the hypersurface of a hypertorus. This is easy to visualize as a cubic crystal lattice. However, the known isotropy of the distribution of matter in space, together with the cosmological principle, impose a different symmetry than this on the actual space of measurement, which is better represented as an equiaxial hypertorus, which is quite unvisualizable. An equiaxial torus can be thought of as a sphere with a duplex surface having zero net curvature; the equiaxial hypertorus is correspondingly as it were a hypersphere with an octuplex hypersurface. The volume of this hypersurface (see table below) is $16\pi^2r^3$. In the cubic lattice model, one lattice interval maps onto one great circle of the hypertorus; this evidently corresponds to the circuit of space, so that the effective 'radius' of space R corresponds to half this circumference. That is to say, $R = \pi r$, where r is the geometrical radius of the hypertorus.

Surfaces and volumes of various figures

r = radius	(Hyper)sphere	Torosphere	(Hyper)torus
Surface	$4\pi r^2$	–	$8\pi r^2$
Volume	$4\pi r^3/3$	–	$8\pi r^3/3$
Hypersurface	$2\pi^2 r^3$	$8\pi^2 r^3$	$16\pi^2 r^3$
Hypervolume	$\pi^2 r^4/2$	$2\pi^2 r^4$	$4\pi^2 r^4$

The volume, therefore, in which the $19.34_{10}55$ gm of particulate matter is dispersed, is $16R^3/\pi$. The value to be taken for R in this formula, denoting the radius of space, is the radius at which the velocity of recession reaches c, beyond which nothing can be observed but up to which everything might, in principle, be seem. This is the Hubble radius. If we take H_0 the Hubble constant for the present epoch, following the best recent estimates, as 55 km s/Mpc, that is to say $1.59_{10}-18\ s^{-1}$, the Hubble radius $R_1 = c/H_0 = 1.885_{10}28$ cm. The volume is then $16(1.885_{10}28)^3/\pi = 3.42_{10}85\ cm^3$. This gives the result that

The mean density of particulate matter in space is $5.66_{10}-30\ gm\ cm^{-3}$. (12.17) C.

This agrees very poorly with most current estimates based on astronomical data, though these have an admitted margin of error great enough to discourage most authors from dogmatic assertions. The weight of evidence is, however, in favour of a density lower, perhaps by two orders of magnitude, than the above estimate. This then is a point against the theory.

However, it is of interest to note that the critical density, computed from relativity theory for the present epoch (using the same value of H_0 as above) to give zero intrinsic curvature to space, is very close to my figure, namely $5.68_{10} - 30$ gm cm^{-3}. If this correspondence is not merely coincidental, the totally independent paths by which the two figures are arrived at implies that one of them incorporates some contant whose value is fixed by their equation. Now there is such a constant, namely the gravitational constant G, which appears in the relativistic argument. It might therefore be claimed that the arguments leading up to our estimate of the density, which rest throughout on the presupposition that space is flat, have a value for determining G, even if for some reason unclear they are not relevant to the actual density of matter observed today.

In fact, the value thus arrived at is

$$G = 6.69_{10} - 8 \text{ cm}^3 \text{ gm s}^2. \qquad (12.12)\text{ C.}$$

This is outside the accepted limits empirically determined, namely $(6.670 \pm .006)_{10} - 8$; but this is hardly surprising in view of the uncertainty in the allowance made for neutrons, which is little more than a guess. We are left puzzled, however, by the low density given by astronomers. This low value is generally said to be supported by the present cosmic abundance of deuterium, which is critically dependent on the density prevailing when it was first formed; this argument is based, however, on the conventional picture of a hot big bang.

12.J. DISCUSSION

In this chapter we have arrived at four quantitative results which can be compared with empirical observation. That for the mass ratio of the proton and electron seems to be correct. The associated estimate of the empirical fine-structure constant is not quite accurate, but being based on a doubtless over-simple model better could not be expected. The value of the gravitational constant G is also wrong, but as it incorporates a quantity only poorly known, the cosmic abundance of neutrons, it is once again as good as (if not better than) we have any reason to expect. Finally, the density of matter in the universe appears to be way out.

It is the case, however, that on the hypertoroidal model the volume contained within the Hubble radius exceeds that of the real universe.

For the latter has a radius $R_0 = ct_0$, where t_0 is the apparent age of the universe, equal to $\frac{2}{3}c/H_0$, and therefore only 2/3 of R_1. Since we are working with a notionally flat space, the volumes contained in these radii are simply proportional to their cubes, showing that the actual number of nucleons that exist is only 8/27 of that estimated, namely $3.42_{10}79$. The assumptions underlying the calculation therefore imply that on the average each nucleon is 'observed' in more than three different 'states' – a possibility already remarked on.

This, however, has no effect on the estimated density, for the reduced number of nucleons are distributed over a correspondingly reduced volume. We are still faced with the evidence for a lower real density, and in consequence for a negatively curved space, all of which is quite contrary to the expectations induced by the present theory. Everything seems consistent, if only space were flat: but it seems that it isn't. We all know what happened to the flat-earth theory. This must be the main argument against my theory from the empirical side.

CHAPTER XIII

GENERAL ASSESSMENT

This final chapter will be given over to an attempt to assess the theory presented in the preceding ones in the light of its correspondence with known facts, or the lack of it, and its capacity for predicting the unknown. Such an assessment could be presented, in the conventional fashion, in the form of an itemized confrontation between theoretical predictions and empirical observations. In fact, I shall not depart radically from this model, though there is a risk that the form may disguise what is going on to some extent. This is because basically we are assessing the evidence for a Hypothesis, (8.1), of an unusual kind. It assumes the existence of a plane of existence different from the physical plane, but manifesting itself by certain constraints which it imposes on the physical. It would be simple, and soon over, if the claim were that such constraints, with their consequences, should explain everything in physics; but the analogy of other plane boundaries, reinforced by common-sense, forbids this.

So we are faced with a messy and open-ended task. Any phenomenon which seems to get no mention in my theory could belong to any of the following types: (a) it is incompatible with some 'constraint' that the theory predicts; (b) it depends in part on contingent factors within the physical plane; (c) it may yet be shown to follow from some aspect of the theory still to be discovered; or (d) it has been overlooked, despite the existence of a simple explanation within the theory. It is possible that this last class is not empty, and of course quite likely that class (c) has some surprises in store; but I fear that most of the obvious gaps will belong to (b), which of course count neither for nor against my Hypothesis. Class (a) too may not be empty, but obviously I believe that that is still arguable, for these cases alone can refute my contention.

13.A. CHARACTERISTICS OF THE THEORY

The enterprise of trying to reduce some of the basic data of physics to consequences of the properties of an underlying plane (in the way that many biological facts can be reduced to physical constraints) has often been considered, but seldom pursued far. Once the nature of this hypothesis is squarely faced, it is obviously going to present considerable difficulties, for it will have little explanatory power unless the constituents of the inchoative plane are radically *unlike* those of the physical plane (not being just some sort of sub-sub-quarks), and unless the inchoative is radically *simpler* than the physical plane.

My thesis clearly satisfies both these requirements; so well, indeed, as to occasion surprise that it imparts any form at all to the physical world. My 'indistinguishables' have required for their investigation the creation of a wholly new mathematics, which makes difficulties for the reader in itself, apart from the inevitably greater difficulty of following the arguments in Part I. But, I feel it right to say, what emerges from Part I is a structure which can be shown to have a recognisable reflection in physical facts – how accurate a one it remains to consider.

It is perhaps inevitable that I may have given the impression here and there that the indistinguishables are *more real* than the more tangible entities dealt with in physical theories. That is, of course, nonsense. Things exist on many levels and planes, and cannot be ranged on a scale of relative realities. You and I, though we are forms persisting with little change as we swim through a long succession of meals, like waves moving over the surface of the sea, are no less real than our food; and if what we eat, reduced to its smallest discernible parts, is seen again as forms 'swimming through' the inchoative substance, it loses nothing in reality by comparison with the indistinguishables to which my hypothesis reduces that substance. We can trace the same path in the opposite direction too, onto the Human plane – but then we soon go beyond the boundaries of science; and the question must therefore be faced, better late than never, whether the theory of indistinguishables ought to count as science at all.

I claim that it does fall within the general purview of the scientific method, for two reasons. First, it is susceptible to mathematical treatment; and second, it leads to inferences which can be compared with observation. But on the one hand the mathematics is admittedly of an unusual kind, and on the other hand there is no account given of how the confirmatory facts are physically related to their alleged causes. This is because we are dealing with the interface between two distinct planes, and may be clarified by the following analogy. It can be predicted from elementary physics that almost all organisms will need to be responsive to the gravitational field in which they live; and so they are. But there is no simple causal relation between the structure and development of the otoliths and suchlike organs which mediate this response and the gravitational field; it is sufficient that they work, and are possible to construct. Similarly, we can expect no causal explanation, from the theory of the inchoative plane, of the physical facts which answer to the constraints deriving from that plane.

13.B. NEW PARADOXES FOR OLD

Science – yes; but physics? – hardly. Physicists have pursued their researches into regions so far removed from direct human experience that what in common-sense terms would be called paradoxes are only to be expected, but some have felt that quantum theory may have grown too accustomed to them to have retained a sense of its own coherence. Does this state of affairs look any better in the light of the present theory, seeing that it is not quite physics anymore?

There are certainly differences; whether they are improvements is a subjective judgment. It is, of course, no peculiarity of my theory that it regards particles as merely hypostatized inferences from events. It is not surprising that divergent mathematical techniques should be applicable to tracing out the effects of things which are themselves no more than creatures of constructive imagination. The wave versus corpuscle paradox loses much of its sting on this view: but that is not new. Once we accept that particles are not tiny billiard-balls but abstract causal relations, we shall be as delighted by scintillations as by interference patterns as evidence in support of our theories.

Some people may find it no less repugnant, however, to find what is persented in one place as a component in a mathematical construction is credited in another with properties such as mass and spin, associated with corpuscles in our imagination. It is of no help to point out, as some philosophers may be tempted to, that the paradox is merely verbal. A food-vessel is a noun, and an auxiliary verb is a grammatical category, and the idea that a given word *can* be read either way, according to context, does not persuade us that there might be *things* which were both foodvessels and auxiliary verbs. But here we are concerned with matters altogether beyond the range of observation, and we have no right to any expectations except for mathematical consistency. Lack of paradox would indeed be very suspicious.

Another notion which may be hard to stomach is that of a 'space' whose 'points' have no ordering among themselves at all, except for a subset of zero measure, nearly all of which are related to what we like to think is real space only through a bizarre statistical relation. When such concepts are presented as mentefacts of a novel theory which we have managed without for so long, it may need more than a few quantitative correspondences to compel conviction. That my disordinate space V can be seen as a limiting case of the multiply-connected 'wormhole' space proposed by Wheeler – an infinitely-connected space in fact – may not be any more welcome in some quarters.

Though it is not strictly a 'paradox', I suspect that a major objection to the theory of indistinguishables, for many physicists, may arise from its implications of finality. It will be obvious that my 'indistinguishables' are not a promising research field: they cannot be observed and have no real properties. But if the theory is accepted, we may be driven to conclude that some of the more challenging parts of fundamental physics may get as satisfactory an explanation from this underlying conceptual level as they are likely to get from more orthodox mathematical investigations. In some circles it is usual to assert that however far we pursue our researches, we shall never reach 'bottom'; I am saying that in some directions we do reach metaphysically impenetrable bedrock. I am *not* saying that this stratum underlies more than a fraction of the territory before us; but the geological metaphor suggests nevertheless that it lies not so very far below the presently known. If high-energy physics is already in contact with 'ultimate unobservables', can the borehole go much deeper?

13.C. A MENDELEEVIAN PRESENTATION

I have in various places pointed out that the rational Sorts, on which the whole application of the theory to fundamental physics rests, form a very recognisable pattern, and have exploited this as a guide to their interpretation. This suggests that a tabular presentation of them might be a useful exercise, on the analogy of the chemists' Periodic Table. This analogy does not survive close scrutiny, which is hardly surprising; but it is I feel more helpful than misleading to emphasize in this way the coherence of the system.

Each horizontal section of the table contains one family of RSs, where successive RSs occupy successive columns. The dual nature of $D3_0 = D2_1$ has occasioned a complication in that an 'object' interpretation is allowed for $D3_0$, which properly belongs rather to the family D2, as is indicated by the step in the horizontal dividing line. The dotted line across the table separates the interpretations claimed as objective and scientific, on the

Family	Type of Interpretation	Dm_0	Dm_1	Dm_2	Dm_3	Dm_4
D0 Locus	Object	$D0_0$ **0** The Void				
D1 Occasions	Dimensions	$D1_0$ **1** Time	and Energy			
	Object	**1** Big Bang	$\underline{D1}_1$ **2** Two Events	$\underline{D1}_2$ **3** 3 Events	$\underline{D1}_3$ **4** 4 Events	$\underline{D1}_4$ **5** Events &c. . . .
D2 Kinds	Components	$D2_0$ **2** of Spin	$D2_1$ **3** of 'Colour' octet	$D2_2$ **7** of Parton 'Flavour'	?	?
	Dimensions	**2** Orientation	and Angular momentum	$\underline{D2}_2$ **10** of full Parton specification		
	Objects	**2** Deg. Freedom of a Boson	**3** Partons in a Boson	**10** Threshold aggregate of ind'ble pions	$\underline{D2}_3$ **137** Threshold aggr. of electrons	$\underline{D2}^*_4$ **$17_{10}37$** Threshold aggr. of nucleons
D3 States	Objects	$D3_0$ **3** Deg. Freedom of a Fermion	$\underline{D2}^*_1$ **5** Partons in a Fermion			
	Dimensions Components	**3** Location of Space vector	and Momentum $\underline{D3}^*_1$ **10** of Energy-momentum tensor 6			
D Percepts	Components	$D\infty_0$ **$2^{\aleph_0}$** of disordinate space				
	Experience	of Perception	$\underline{D\infty}_1$ Awareness of Perception	$\underline{D\infty}_2$ Awareness of awareness	$\underline{D\infty}_3$ Etc. . . .	
D Totality	Experience	D1 of the Whole				

left, from those of a subjective character mentioned in the text. The entries under 'type of interpretation' are explained in the course of Chapters IX–XI. Those who object to 'degrees of freedom' being entered as 'objects' may read 'quarks' instead; the correspondence is not exact (see Section 11.F), and I have therefore avoided this term in the table.

There are two blank spaces in the table, for 'component' interpretations of $\underline{D2}_3$ and $\underline{D2}^*_4$. I leave undecided whether there is really anything to be discovered here, or whether these are simply to be taken as presence-or-absence entries relating to the thresholds given in previous page.

13.D. SUMMARY OF EVIDENCE

In this section I give in tabular form a collection of the conclusions we have arrived at, together with a note on where in the preceding text the matter is dealt with, and what kind of confirmation, if any, the conclusion receives. The following notations are employed for this purpose:

(?) =	doubtful assessment	Q =	qualitative agreement
FU =	facts unknown	QC	with common knowledge
IC =	indirectly confirmed	QO	with observation
Nx =	numerical agreement to x decimal places	QT	with accepted theory

X = conflicts with evidence

Our conclusions are

1 : There exists a Void, populated by discrete observable Events (9.1)

1.1 : which constitute a denumerably infinite Set

1.2 : in which we can distinguish one ordering corresponding to 'time' and three independent twin orderings as descriptive of 'space' (9.6) QC

1.3 : 'time' having no upper bound, whereas 'space' is finite (9.8) (?)

2: To an observer

2.1: the Void appears as an extended dense 4-space

2.1.1: in which the Events are located, and confer on regions containing more than one of them the above ordering, but

2.1.2: in any smaller region of which points can be specified only subject to 'disordinate statistics'. (10.1) IC

2.2: Measurements can be made on Events in this 4-space

2.2.1: subject to an irreducible error affecting pairs of simultaneous measurements of quantities whose product has the dimensions of action; and those (11.1) QT

2.2.2: quantities, whose action-complements form or contribute to a system having up to three degrees of freedom, are conserved in any closed region, (11.2)

2.2.3: from which a system of metrics for the space can be derived, (10.4) QT

2.3: under which the overall configuration of the space is that of a (convex figure tangent at the apex to a) right hypercone whose base is the hypersurface of an equiaxial hypertorus intersecting the conical axis at the location of the observer. (10.12) FU*

3: The Events thus observed are

3.1: of various kinds, whose relative frequencies are dependent on time in any closed region (10.7) QC

3.2 : and are individually

3.2.1 : not necessarily invariant under reversal of chirality alone (10.13) QO

3.2.2 : but are invariant under time-reversal (which entails that of chirality) (10.8) QO

4 : Causal relationships between Events can be hypostatized as particles, which are subject to

4.1 : an upper limit on their velocity (observable as the velocity of light *in vacuo*) (10.9) QT

4.2 : which is independent of the motion of the observer (10.9a) QT

4.3 : and is equal to the initial value of the rate of increase of the extent of space (10.10) QT

5 : Each Event can then be characterized by the particles participating in it, each of which is specifiable

5.1 : as to kind, as either

5.1.1 : a boson, which is an aggregate of three partons, each specified by a descriptor of ten components, of which (11.2)

5.1.1.1 : three underlie 'colour', of which those predicated of any two partons determine the third, and QT

5.1.1.2 : seven underlie 'flavour', of which those predicated of any one parton determine the rest; or (11.6) QT

5.1.2 : a fermion, which is an aggregate of five partons, similarly specified, such that ?

5.1.2.1 : the 'colour' predicated of any three partons determines the remainder, and QT

5.1.2.2: the 'flavour' predicated of any three partons determines the rest

5.2: and as to state, by the ten components of the energy-momentum tensor (12.1) QT

5.3: and of which two kinds having finite mass are stable, the ratio of their masses being 1836.1515 ± 0.00005 (12.9) N8

5.4: and which may interact in three ways, characterized by stability-thresholds of the particles involved when concentrated within the Compton radius of one of them, namely

5.4.1: the strong interaction, possibly with a threshold of 14 pion-pairs (11.8.5) N(?)

5.4.2: the electromagnetic interaction, with a fission threshold of 137 electron-positron pairs and a coupling-constant 0.0072972 appr. (11.8.4) N3 N5*

5.4.3: the gravitational interaction, with a collapse threshold of $17_{10}37$ nucleons and a coupling constant of $5.88_{10} - 38$. (11.8.3) N3

6: The density of Events over parts of space

6.1: tends, over sufficiently large regions, towards a statistically uniform value (11.5) QT

6.2: which at the present epoch, assuming the Hubble constant to be 55 km s Mpc, amounts to about $5.66(\pm .05)\,10^{-30}$ gm cm^3 of particulate matter, whence (12.11) X

6.2.1: the deceleration parameter q_0 of the Einstein–deSitter cosmological model should be close to 1/2, and

6.2.2: the gravitational constant, assuming the truth of (6.2.1), is $G = 6.69(\pm 0.06)$ $10^{-8}\ cm^3\ gm\ s^2$ (12.12) N2

13.E. ASSESSMENT OF THE RESULTS

These results are the end-product of the process of fitting the raw data tabulated in (7.21) to a set of interpretations which are strongly constrained by the principle of coherence. In a few cases (marked with an asterisk in the rightmost column) an element of conjecture not inferred from the theory is involved; of these, 2.3 is presupposed as true by 6.2. Two questions occur here: Are the results correct to within the limits of error in the calculations and of the relevant empirical data? and, if so, Why? The first of these will be examined below; the second question is a physical one, and because the theory of indistinguishables concerns another plane than the physical, it provides no answer. It might be thought that, strictly understood, the theory should preclude there *being* an answer to the 'why?'. This I do not accept: the way the results are set out above does not give an accurate impression of the *number* of constraints which, if the theory is correct, the physical plane is subject to as a consequence of its dependence on the inchoative plane; this is difficult to estimate, but is probably in the range 8 to 12. The number of assessable results listed above is 24. The number of physical constants not explained already in terms of others is much greater still. There is therefore plenty of room for answering 'why?' questions about them, even about the small subset appearing in Section 14.D, in terms of strictly physical theory.

It is obvious that the results are of very unequal value for confirming or refuting the theory. The qualitative agreements may be accepted only as confirming an assessment more solidly based, though their number and diversity of content may legitimately cause surprise if other evidence is unfavourable. Some of them, however, are not beyond the possibility of losing their 'QT' status: current rapid developments in the QCD field may call in question those in 5.1, though the formal basis of this theory is hardly yet sufficiently clarified for a final judgment. In particular, the implication that leptons have a similar constitution to hadrons will not be accepted be many – which has caused me to put a ? against entry 5.1.2.

Apart from this case, there is really only one item which gets a seeming refutation, namely 6.2; the query on 1.3 arises from this. Estimates of the mean density of the universe vary enormously, but direct astronomical observation currently suggests a figure one or two orders of magnitude less than mine. Those concerned with these observations seem generally willing to admit a lack of confidence in them, and it is permissible to believe that the 'X' assigned to (6.2) may one day be removed. But as of now, it stands: so what? It is true that the argument leading to my estimate of the density is by no means rigorous, and may be quite misconceived; on this ground then I might dodge the issue, but for one point. This is, that my figure is within present limits of error equal to that required to make the universe exactly 'flat', which is what the rest of theory would predict; such an internal consistency is hardly probable if my calculation were quite wrong. I have therefore added the entry 6.2.2 on this basis (the error limits mentioned here arise from the uncertainty in the neutron correction discussed in Section 12.6). Perhaps the world *is* flat?

Turning now to the successes, these will be found in 5.3 and 5.4. Apart from 5.4.1, to which little weight can be attached at present, there are four of these, all within experimental limits of error except the fine-structure constant, the calculation of which has involved some pretty crude geometry, acknowledged by the asterisk. What is the weight to be attached to four such correct 'predictions'? It is at this point that the question, raised earlier, of *why* they come out right becomes insistent. Would they cease to support the present theory if they were derived with equal success from some strictly physical argument? Not necessarily, unless it could be shown that there are unacknowledged physical determinants hidden in my calculations (this could only apply to 5.3); it is obvious that no 'constraints' emanating from the inchoative plane could conflict with the internal consistency of a sound physical theory. It might rather be that that theory already contained a hidden input representing such a constraint.

I therefore feel it legitimate to conclude that we have in these numerical agreements a strong support for the essential correctness of the Hypothesis (8.1); a support corroborated by the numerous qualitative agreements listed in other parts of Section 14.D. Sufficiently so at least to warrant its public presentation and wider discussion.

13.F. LIMITS OF INTERPRETATION

It has been already emphasized that the Theory of Indistinguishables is not a physical theory. It purports to predict constraints, arising from the internal properties of a hypothetical inchoative plane, which are expected to appear as characteristics of the physical plane for which there may be no adequate explanation within that plane. The results listed in Section 14.D are all candidates for consideration as such constraints. It is not to be expected that they will all appear as physically independent; but a complete physical theory, when this in due course emerges, must incorporate certain axioms and unexplained inputs, unless it can be claimed as an *a priori* theory, which I assume is no longer expected by anyone. These unexplained inputs could be a *subset* of the results of Section 14.D; if my theory is correct, they will be so.

This indicates that the Theory of Indistinguishables might have a further role in identifying what things are legitimately in need of a purely physical explanation. In brief, any phenomenon which is *not* predictable from my theory must be accorded this status. Obviously, not to be predictable is not the same as not to have been predicted, and just to scan the table of results for gaps will not necessarily suffice for this purpose. I cannot claim to have exhausted the explanatory power of my theory, the more so as my knowledge of physical theory is insufficient for that task. But no one will doubt that there are many gaps.

One of these is in its conception neatly complementary to my theory. The 'bootstrap' theory of hadrons, associated with Geoffrey Chew [5] and his associates, aims to predict the various species of hadrons, and hopefully their masses and lifetimes, from considerations of mutual compatability applied to the set of all possible S-matrices. This is of course a truly physical theory (its apriorism excepted), and aims towards an important topic which is completely left out of my account. It is too early to speculate on the success of this venture, but if it comes off it will be a welcome complement to what we have here.

Another obvious gap is my failure to account for the actual values assigned to 'quantum numbers'. These may be explained by the combinatorial hierarchy theory [4], which closely parallels the theory of indistinguishables but proceeds entirely within physical concepts. Such a theory is to be expected, if there is any truth in my hypothesis, for without it the absence of a physical 'dimension', though perhaps explainable, would be legitimately puzzling. This could also provide for physical interrelations between some of the items listed among my results.

13.G. GENERAL CONCLUSIONS

Let us now try to draw the threads together, with a view to determining whether the evidence is sufficient to sustain the Hypothesis stated in

Chapter VIII, and if so what follows from it for physical theory in the wider sense.

It is necessary first to remember that the evidence adduced in support has not uniformly been favourable, for item 6.2 runs counter to the best opinion current on the matter of the mean density of matter in the universe. In view of the uncertainty of this evidence, we are not bound to abandon the theory; but it may yet be that what it describes is an ideal, self-consistent, universe, which is not fully realized in actuality, for some undiscovered reason. Even if this were so, it would not be worthless; but it is a lame conclusion nevertheless, and I am inclined therefore to make less of this reverse than perhaps I ought to.

While on cosmological topics, I ought to make some reference to the implications of the theory in relation to events just after the 'big bang'. There is good ground, if the theory can be in general accepted, for reconsidering the whole question of cosmogony. It is after all one of my stated conclusions (consequent on 3.1) that processes involving many events are not reversible in time; this being so, it is not legitimate to imagine the beginning of the universe as if it were a super-black-hole run backwards. Whether any suggested alternative makes any sense cannot be considered at this point: but it is at least clear that any theory of indistinguishables must question any picture of the first few seconds which assumes that everything possesses separate identity from the very first. There is radical rethinking to be done on this topic, which is unlikely to add to the appeal of my ideas for those who have committed so much effort to the present conception.

Perhaps the most striking feature of the theory, on surveying the table of reults, is that a surprising variety of conclusions emerge from a seemingly very inadequate set of initial assumptions. This, I suggest, is simply in line with what one ought to expect from a theory which deliberately takes its stand on an assumed pre-physical inchoative plane of existence. If there is such a plane, it cannot be very complicated (or it would need another to explain it, which would effectively refute the original idea should it turn out to be the case), and yet it must constrain the physical plane in many ways (or, again, fail in its purpose). That is to say, an exhibition of *multum ex paruo* is a necessary mark of success in a theory of this kind, and is only remarkable if the case is first made that it has in fact failed.

The role of the observer in the logical validation of the theory has been touched on only briefly, and may be a point of contention. I have been able to sidestep this by the formulation of my hypothesis without explicit mention of observers; but only at the cost of producing an impoverished universe, for what does not go in in Chapter VIII also does not come out in Chapter XIII. What we have here is a finite universe, and therefore one inhospitable to speculation on completed infinities, which are presupposed in my claim that there are rational infinite Sorts other than $D\infty_0$. This suffices to provide a basis for the space V, but tells us nothing about the process of observation. That only enters the picture if we imagine that our world contains beings for whom the whole family $D\infty$ makes some kind of sense: which it does, of course, and observation can indeed go forward. But to arrive at this point we have to add to the universe of Section 14.D an extra ingredient, a second 'moment of creation', which takes us beyond the proper bounds of scientific theory. It could therefore be argued that my theory is not fully self-consistent unless extended in this manner, on the grounds that it does not account for the evidence which supports it having become available. I dismiss this criticism as metaphysical and so out of order, though I do not in fact deny its truth.

On a more mundane level, however, the theory stands up well. It does not provide physical explanations for anything; but it does provide a different kind of 'explanation' for some of the basic characteristics of the physical world, including several of the dimensionless ratios which have hitherto resisted explanation within physics itself. One could take the line that it is unreasonable to *expect* for these things a physical explanation (where are the inputs for the argument ?); and even if one pushes to the utmost the observation that if they were even slightly different from what they are the universe would be unable to provide a home for mankind, the Creator would emerge as ingenious rather than as numinous. The Theory of Indistinguishables, if it can from its seemingly meagre premises compute their values correctly, closes this line of argument.

Indeed, too much seems to be right to allow one to dismiss the theory out of hand, even though obviously all is not yet right. And it has yet to be proved that physics can be consistently erected on *any* subset of the theory's conclusions, which if it is not the case demolishes the whole edifice. But, on the results so far, I feel it right to suspend judgment, and to commend the theory to the criticism of the informed public. It may yet fail: but it may also prove to be a useful step in the long task of understanding the nature of the world we live in.

REFERENCES

[1] Ramsey, F. P.: *The Foundations of Mathematics, and Other Logical Essays* (ed. by R. B. Braithwaite), International Library of Psychology, by London, 1931.
[2] Whitehead A. N. and Russell, B.: *Principia Mathematica*, Cambridge University Press, Cambridge, 1914.
[3] Bourbaki, N.: *Elements of Mathematics*, Vol. I, Adiwes, Paris, 1966.
[4] Bastin, T., Noyes, H. P., Amson, J., and Kilmister C. W.: 'On the Physical Interpretation and Mathematical Structure of the Combinatorial Hierarchy', *Journal of Theoretical Physics* **18** (1979), 445.
[5] Chew, G. F.: '"Bootstrap": A Scientific Idea?', *Science* **161** (1968), 762.

INDEX OF TERMS DEFINED

Each reference is an initial substring of Chapter – Section – Statement No. – Entry No. – **plus Page No.**

Examples: 1G.3N1.3 = Chapter I, Section 1G, statement 1.3, entry in this statement headed 'N1.3'.

4 = Chapter IV, introductory paragraphs.

INDEX OF SYMBOLS

The order of listing the symbols is alphabetic–numerical–ideographic; variants of one letter (incl. u.c./l.c.) or numeral are placed together in historical order. Ideographs are ordered as I think best.

Relators

Identifiers (not of classes)

Classes (incl. Functors denoting such)

Functors (other than relators)

GPSR Compliance

The European Union's (EU) General Product Safety Regulation (GPSR) is a set of rules that requires consumer products to be safe and our obligations to ensure this.

If you have any concerns about our products, you can contact us on ProductSafety@springernature.com

In case Publisher is established outside the EU, the EU authorized representative is:

Springer Nature Customer Service Center GmbH
Europaplatz 3
69115 Heidelberg, Germany

Batch number: 10607222

Printed by Printforce, the Netherlands